Basic Robotics Concepts
by

International Standard Book Number: 0-672-21952-2
Library of Congress Catalog Card Number: 83-60173

Edited by: *Arlet Pryor*
Illustrated by: *T. R. Emrick*

Printed in the United States of America.

Preface

Of all the many innovations of our time, the development of intelligent machines promises to be among the most significant. The field of intelligent machines has many facets, including computer-aided design (CAD), computer-aided manufacturing (CAM), automatic process control, general automation, and robotics. Although all these fields overlap and are interdependent, none is more visible than robotics.

The very word *robot* calls up strong feelings in almost everyone, but these feelings vary immensely between individuals. To the engineer or technician, robots offer the opportunity to create animated machines that seem almost "alive." This possibility is enough to excite even the most sedate technocrat. To the manufacturer, the robot may seem to be the ultimate weapon in a lifelong struggle with organized labor. To the laborer on the assembly line, the word *robot* may loom overhead like a dark shadow.

Fortunately, things are not as one sided as these perceptions would indicate. For one thing, the manufacturer will need markets for all of the goods that the robots create, and the present blue collar work force represents a very significant part of such a marketplace that cannot simply be cast aside. Furthermore, the assembly-line worker will be thrust out of what is often a boring and degrading job in the first place. With a bit of effort, workers can turn this into a golden opportunity. The trick for them will be to prepare for the change as early as possible, instead of simply waiting for the ax to fall. With a modest amount of research, such individuals could even be in a position to work with the robots that have replaced them at their old jobs.

Robots offer other benefits as well. The average consumer will be able to purchase robot-built goods that are less expensive and of consistently better quality than those built by hand. The chance of buying a "lemon" automobile will be greatly reduced because robots work as carefully on Monday and Friday as they do during the rest of the week. For that matter, they work weekends as well.

To fully realize the potential behind robots, consider the following fact: Virtually all the great civilizations of the past were based, at least in part, on slavery. The ancient Romans, Greeks, and Egyptians all used slaves to free their citizens from the trivial and monotonous work of daily existence. Unfortunately, the enslaving of other peoples brings with it a high moral price. Now there is a way for people to have slaves without injustice. In fact, these slaves could eliminate many of the economic injustices that exist today.

Freed by robots from the boredom of dead-end jobs, more people will have time for the higher pursuits. Art, philosophy, music, science, and all forms of learning can be given more attention. Improving the environment, making our cities more pleasant places, and reducing crime can all be given higher priorities. The long-term effect of robots will therefore be largely dependent on how responsibly we, and our leaders, adapt our social structure to their use. But whether we use them wisely or foolishly, *the robots are coming.*

ACKNOWLEDGMENTS

I wish to extend my sincere thanks to Dr. Christopher Titus of the Blacksburg Group for his guidance and editorial assistance with this book. His painstaking efforts have been of great benefit to both the book and to me.

I would also like to thank Mr. Robert Pharr of Roanoke, Virginia, whose mechanical expertise brought the mobile robot "Kludge" from a concept to a reality. Thanks must also go to my technician, Mr. William Grady Spiegel, for his support in breadboarding many of the circuits associated with Kludge and other systems discussed in this book.

Finally, I would like to thank the many engineers, researchers, and other kind people who helped make this book possible by suffering politely through my stupid questions and by supplying information, photographs, and fresh ideas. Among these people, special thanks must go to Dr. Hans Moravec of Carnegie-Mellon University, Dr. R.B. McGhee of Ohio State University, and Hurley Gill of Inland Motors.

JOHN M. HOLLAND

Contents

CHAPTER 4

The Vidicon Camera—Plumbicons and Other Relatives of the Vidicon—Composite Video—Charge Coupled Devices (CCDs)—Other Solid-State Camera Technology—Optical Considerations—Digitizers and Preprocessors—Full-Frame Image Processing—Two-Dimensional Processing—Recognizing the Closed-Edge Image—Area, Moments, and the Axis of Minimum Inertia—The SRI Algorithm and Moment Invariants—Color Vision—Three-Dimensional Video Processing and Range Determination—Structured Light Vision Systems—The Vision System as an Integrated Part of the Robot—Conclusions

APPENDIX A

Motor Equations for Chapter 1—Gear and Actuator Equations—Torque Conversion Table—Rotary Inertia Conversion Table—Mobility Equations for Chapter 3—Equations for Chapter 4

APPENDIX B

Program CG.BAS for Determining the Center of Gravity—Sample Run of Program CG.BAS—Program STABLE.BAS for Determining the Limits of Stability—Sample Run of STABLE.BAS

Introduction

This book has been written with the hope that it will be useful to a wide range of readers. The book addresses the four most central subjects of robotics: motion control, manipulators, mobility, and vision. A host of other subjects interplay with the field of robotics to a lesser degree, including active vision, navigation, communications, and speech. The proper treatment of all these subjects is beyond the scope of a single volume, so this book will serve as an in-depth introduction to the basic concepts behind robots.

The subjects covered will be explained in a depth that should give the technically oriented reader a basis for evaluating potential robotic projects or for continuing research. For those readers who require mathematical formulas, these have been compiled in Appendix A. It is hoped that this placement will simplify referring to these formulas as well as making the chapters more readable. This structure has required that the concepts be explained in words rather than equations. Once they are understood, most of these concepts will seem remarkably simple.

It is also hoped that robotics hobbyists will find the book useful, even though the full implementation of some concepts may be beyond their financial resources. In fact, operating on a limited budget can be the catalyst for significant innovations. One has only to look at the early microprocessor hobbyists to see the importance of such individuals. Although the home computer market is now filled with equipment bearing names such as IBM, DEC, Xerox, and Hewlett-Packard, these were not the innovators. The innovators were hobbyists, enthusiasts, and entrepreneurs, who formed fledgling companies like Apple Computer, Inc., Imsai, Altair, and others. Some of these individuals had more formal training than others, but they all had enthusiasm and a taste for technical adventure.

At the other extreme, the manager or worker wishing to become familiar with the subject should find the book useful. Those readers who require only an overview of the various topics covered may wish to skim through the chapters, by beginning at each subtopic, reading until the level becomes more technical than is needed and then skipping to the next subtopic. Particular notice should also be taken of the summary at the end of major subjects. In this way, an insight to the capabilities and pitfalls of various robotic systems can be gained.

A SHORT HISTORY [1]

Robotics is not a single technology but the combination of many technologies. Some of these fields were mature long before others. By the seventeenth century, mechanical technology was already advanced enough that craftsmen were able to fabricate mechanical robots. Some of these robots could actually draw sketches and do other remarkably complicated functions. The limiting factor with these robots was the programming. The use of cams and cogs as a means of programming made the devices temperamental, hard to program, and expensive. Additionally, it is difficult to fabricate sensory functions into purely mechanical systems.

During the 1950s, engineers began applying electronic controls to robotic devices. During this period, computers were extremely expensive and inefficient. With an instruction cycle rate of 50,000 cycles per second, they were a factor of one hundred slower than the average home computer of today. This factor alone made these computers inappropriate for real-time control applications. As a result, analog techniques were generally used.

One of the first practical robots was introduced in 1955 by the Planet Corp. The robot was used for handling hot castings and was reprogrammable. Also during this period Joseph Engelberger and George Devol met and began working on the problem of a practical industrial robot. In 1961, their new company (Unimation, Inc.) put its first industrial robot on line in a die-casting application.

The science of linear control systems was already fairly well advanced by the late 1950s, and its choice for robotic systems was the least objectionable of the alternatives available. Interestingly, the fatal shortcoming of these systems was the same as for their mechanical predecessors—programming. Although analog systems could accurately accomplish closed-loop control, they could not be sequenced easily through a series of operations.

The early analog robots had their shortcomings. The most common approach to programming these analog systems was to sequence

8

the input of the position servo amplifiers through a number of voltage steps that corresponded to the desired end points of path segments. The sequencing was done by rotary stepping relays (developed for the telephone industry), and voltages were programmed by an array of potentiometers. This system was very limited in the number of paths that could be programmed. Additionally, conditional and relative motion was almost impossible. These limitations prevented the analog robots from reaching any degree of popularity.

The development of a practical robotic arm, which could move a tool accurately along an arbitrary line in space, would have to wait for the development of the minicomputer. Such an arm is referred to now as being *fully articulated*. It was, however, possible to develop an arm that resembled a modern articulated arm, but whose control was limited to simple sequences.

The control of these nonarticulated arms was done by simply driving the actuators of the arm until stops were reached. These stops were usually adjustable limit switches. The arms could perform simple, highly repetitive functions such as moving parts from one position to another. Unfortunately, such systems do not have the ability to significantly change their movement pattern in response to external conditions. Despite this restriction, nonarticulated arms were, and still are, widely used in industry. With the addition of better controllers such as *PCs (programmable controllers)*, these arms can be interfaced with, and synchronized to, conveyors and robots. For the purposes of this book, these machines will not be regarded as robots and will not be discussed in any depth.

During the period between 1961 and the advent of inexpensive minicomputers (and later microcomputers), Unimation, Inc., was the only company in the field to continue a strong program of research and development. (By the middle 1970s, Unimation, Inc., was producing practical, reliable robots in substantial numbers.)

ENTER THE JAPANESE

The importance of industrial robots was becoming apparent by 1978, and the Japanese were already involved in what has now become almost a "love affair" with this exciting new technology. Startled by the enthusiasm of the Japanese, the western press generated a series of reports showing Japan as having an enormous lead in the number of robots installed. Interestingly, much of this early disparity was because the Japanese counted the nonarticulated arms as robots, whereas the Americans did not. This fact should not, however, lull the American manufacturer into underestimating the commitment to robots that has permeated the entire Japanese culture. In general, Japanese maufacturers assure their workers that the robots will not cost them their employment. Instead, the workers are retrained for more interesting and demanding tasks, and any reduction in the work force is accomplished by a reduction in hiring.

Such a commitment to the long-term interests of the employees is contrary to the near-term management philosophy of most American industry. Indeed, even the expenditure of significant sums of money for capital improvements (such as the incorporation of robots) was discouraged by the system that became dominant during the 1960s and 1970s. Under this system, top executives increased their upward mobility by jumping from one organization to another. To compound the problems, most companies rewarded performance by a quarterly or yearly bonus system. This bonus system was linked to the profit of the company over that period and did not reflect the future prospects in any way. For a manager who intended to move on in a year or two, a long-term investment in either the plant or the employees would seem foolish in the extreme.

In addition to this nearsighted philosophy, most western societies have become more impersonal in recent decades. Markets and labor forces are now thought of as great homogeneous masses instead of collections of individuals. When order input drops, most companies simply lay off workers. The whole process is visualized as opening a valve and letting the excess labor flow back into the local labor pool. Such an attitude has been encouraged by labor unions that insist that their members be treated as interchangeable cogs. The effect of all of this on the average worker has been to cause total alienation.

It seems likely that this system is at least partially to blame for the fact that American workers have relatively negative feelings about robots, while the Japanese have equally strong positive feelings. Ironically, the robots themselves offer an opportunity for American industry to break this spiral.

THE FUTURE

There is a growing tendency for those in the industrial sector to regard the word *robot* as synonymous with the popular articulated arms. In fact, these machines are only one piece in the factory of the future. Mobile robots are only beginning to be considered for industrial applications although driverless carts and trains have been in service for years. Multiple-arm robots and other unconventional types are becoming available as robots move into new fields of application.

Robots are finding application outside the factory as well. Practical robots are at the verge of entering the agricultural and mining industries, and the military is sponsoring a good deal of basic research for its applications. What we are experiencing now is the birth of a vast and exciting new industry, and ample opportunities are available for anyone who will reach out and grab them. To begin to understand it all, let us begin with the basics.

References

1. "Technology Growth Markets and Opportunities," Volume 1, No. 4. Creative Strategies International, 4340 Stevens Creek Rd., San Jose, CA 95129 (408) 249-7550.

Chapter 1

Motors
and Methods

The capacity for self-directed motion is perhaps the most fundamental characteristic that separates a robot from other types of machinery. Besides this capability, a robot may or may not have other capabilities (such as ranging and vision). The ability to produce accurately controlled motion is not only critical to the motion of the robot, but it is also important to the systems that provide the robot with many of these other capabilities. For this reason, motors and methods of motion control will be discussed first.

The science of motors and motion control is far from simple, and entire volumes have been written about single aspects of the subject. No attempt will be made here to cover the subject in any depth. Instead, an overview will be presented that should give a general feeling for the components that are available, and how they may be applied to robotic applications. For those wishing to pursue specific topics further, several excellent sources of information are listed in the references at the end of the chapter. Additionally, Appendix A contains many of the mathematical formulas associated with servo-system design.

TYPES OF SYSTEMS

The three most common types of motion control systems are hydraulic, pneumatic, and electrical. Each of these systems has advantages and disadvantages. Hydraulic systems, for example, can deliver a great deal of power with very little weight at the actuator. This characteristic makes a hydraulic actuator very attractive at the end of a long manipulator arm (where inertia is critical to performance). Unfortunately, these systems have a tendency to leak hydraulic fluid. They are also somewhat more difficult to control than electrical systems. Additionally, if the primary energy source is electrical (ac power line or batteries), an electrically powered hydraulic pump is required. This additional conversion of energy causes a significant loss of efficiency and reliability. On the other hand, for a robot powered by an internal combustion engine, a hydraulic system might be more efficient than an electrical servo system.

Pneumatic systems are similar in structure to hydraulic systems, except that the oil is replaced by a compressible gas (usually air) as the medium of energy transmission. Generally, pneumatic systems are cleaner and lighter than hydraulic systems. The compressibility of the transmission medium brings both advantages and disadvantages. When used with high-inertia loads, pneumatic systems tend to require special damping to prevent oscillation. This is a direct result of the spring in the medium. Compressibility also provides a degree of *passive compliance.* Compliance is essentially the characteristic of a system to trade-off position for force (see also Chapter 3). For this reason, pneumatic systems are often used in grasping mechanisms (see Chapter 2). When pneumatic systems are used for joint actuation, an air motor is usually used.

Both hydraulic and pneumatic systems have additional disadvantages. Conventional hydraulic and pneumatic systems utilize valves to affect control, and this utilization often leads to a certain amount of position hysteresis as the feed lines to the cylinders change from positive to negative pressure. Certain techniques of control (such as mounting the control valves at the actuator) may minimize this tendency, but there is still another disadvantage. Because hydraulic and pneumatic systems require such items as compressors, pumps, accumulators, and tanks, they tend to absorb larger areas of valuable floor space than do electrical systems. Even with all of these problems, the advantages offered by hydraulic and pneumatic systems continue to dictate their use in many applications. More will be said about these systems later in this chapter.

Electrical systems come in a very wide and growing variety of forms. Besides the trade-offs that must be considered between electrical systems and other systems, important differences exist between the various components of electrical systems.

ELECTRICAL MOTORS

All practical electrical motors are based on the fact that an electrical current flowing in a conductor generates a magnetic field.

Various types of motors use this principle in different ways to create motion. Since robotic systems almost always require accurate control of either the velocity or position of a motor, synchronous ac motors will not be discussed.

Conventional Commutated Motors

One of the oldest and most widely used types of dc motor is the *conventional commutated motor.* As shown in Fig. 1-1, the construction of this type of motor features a rotating armature that consists of several coil windings. In real motors (as opposed to the simplified drawing of Fig. 1-1), these windings are laid in slots in the rotor. The rotor is usually constructed using thin laminated metal disks to reduce eddy currents. The armature is powered through a commutator and is surrounded by a stationary stator. The stator (also referred to as the field) may consist of one or more permanent or electric magnets.

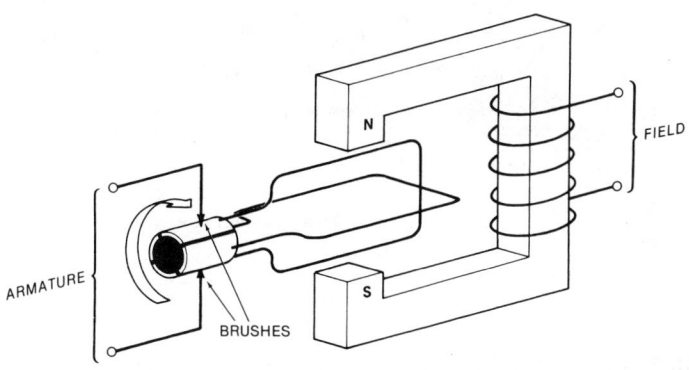

Fig. 1-1. Conventional commutated dc motor.

The commutator is connected to the rotating armature in such a way as to switch the direction of the current in the windings of the armature so that the magnet field of the armature is not aligned with that of the stator winding. The opposite poles of the armature and stator will therefore be attracted and the armature will rotate. As the armature rotates toward alignment with the stator, the commutator will constantly switch the current in the armature so that the rotation continues. The effect of this switching action is to cause the magnetic field of the armature to remain fixed in space (within the magnitude of the commutation angle).[1] The angle between the stator and armature magnetic flux patterns is referred to as the brush or neutral angle. If a dc motor is to operate bidirectionally, this angle must be approximately 90°. In unidirectional motor applications, some tailoring of the motor's characteristic speed/torque relationship may be accomplished by changing this angle.

The commutator may consist of a simple set of carbon brushes running on a segmented contact ring, or it might consist of some other type of position sensor and a solid-state driver. Motors as in Fig. 1-2 have recently become available with magnetic (Hall-effect) commutators. These commutators have a longer life expectancy than do

Courtesy Inland Motors, Industrial Drives Division, Kollmorgen Corp.

Fig. 1-2. Cutaway view of conventionally commutated samarium-cobalt permanent magnet dc motor.

brushes, and they eliminate sparking. In applications where combustible or explosive elements may be present, such commutators can be very valuable. Several manufacturers of spray painting robots have begun using this type of motor for this reason. Noncontact brushes can also greatly reduce the *rfi (radio-frequency interference)* generated by the motor. This improvement has strong implications for mobile robots that utilize radio telemetry.

Obviously, the advantages of eliminating contact brushes would be largely lost if the armature has to be powered by slip rings. In order to prevent the need to provide power to the rotor, these motors usually use what is referred to as an *inside-out construction.*

Inside-Out Commutated Motors

The inside-out commutated motor construction shown in Fig. 1-3 features a permanent-magnet rotor and fixed, commutated windings to produce a rotating magnetic field. With these motors, the heat generated in the windings can be transferred by thermal conduction to the outside case of the motor. In the higher-power motors, this avoids the need to blow air around the armature, allowing the motor to be sealed against dirt. Alternatively, air *is* sometimes blown through the motor resulting in much greater heat transfer than in conventional motors. Lower rotor inertias can also be obtained using this construction. For these reasons, even many of the more powerful contact brush-type motors are now being constructed with the inside-out approach.[M2] More will be said about these motors in the section on permanent-magnet motors.

Wound Field Motors

Wound field motors contain one or more field windings on a stator frame to produce a stationary magnetic field. As the armature rotates in this field, the effect is for it to act as a generator and develop a back electromotive force (emf). This voltage bucks the drive current and thus limits the speed of the motor. For any given drive voltage, the motor will reach a velocity at which the back emf (combined with the brush and motor IR voltages) cancels the drive emf. If the current in the field winding is reduced, less back emf will be generated, and an unloaded motor will run *faster.* The torque capability of the motor will also be reduced as the field is diminished.

As a motor is loaded, the rotational velocity of the armature will be reduced, and less back emf will be generated. This loss of *bucking voltage* means that the effective voltage across the windings of the armature is increased, and thus the current will rise. These characteristic interactions between the field and the armature can be used to control the relationship between speed and torque in a motor.

Parallel Motors

The most obvious technique for energizing the armature and field of a motor is to simply connect both the field and the armature (through the commutator) to the drive voltage. Since the two wind-

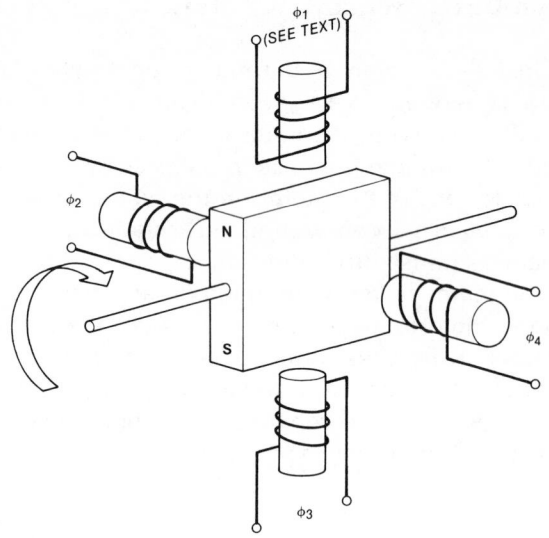

Note: Actual motors have many magnetic poles

Fig. 1-3. Inside-out motor.

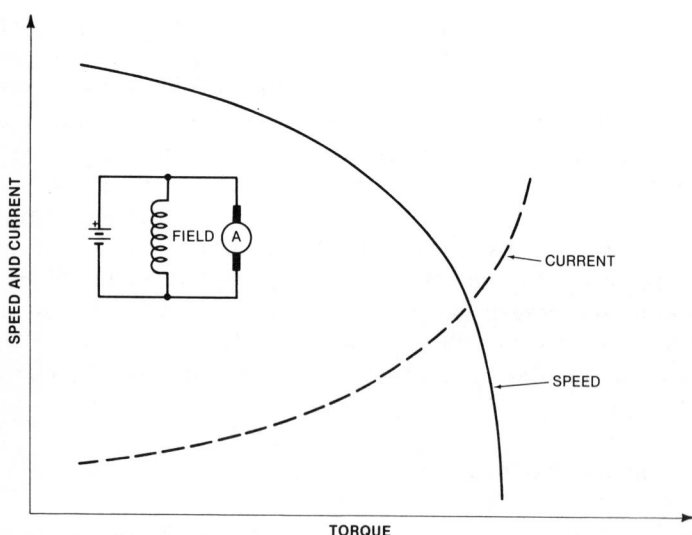

Fig. 1-4. Speed-torque curve for a typical shunt motor.

ings are connected in parallel, this configuration is called a *parallel motor.* The curve for torque versus speed in a parallel motor is given in Fig. 1-4.[1] Notice that at the low-speed end of the curve, achieving additional torque requires a more significant loss of speed, which is to say that the torque-speed curve goes nonlinear.

Series Motors

If the field winding is connected in series with the armature winding, the relationship between torque and speed is much different as shown in Fig. 1-5. In a series motor, the field winding carries

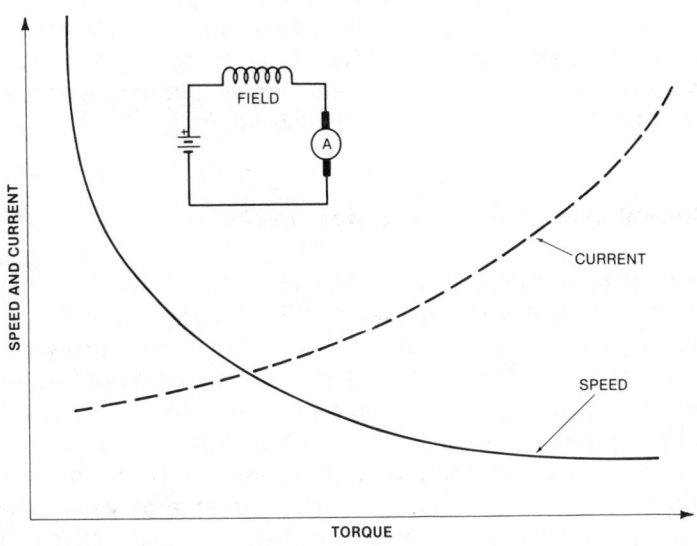

Fig. 1-5. Speed-torque curve for a typical series motor.

the full current of the armature, and thus is usually composed of many fewer turns of much heavier wire than the field winding of a parallel motor. When a series motor is at low speed (during start-up or as a result of heavy loading), the armature will develop very little back emf, and the armature current will rise accordingly. Since the armature current flows through the field winding, the flux density (magnetic intensity) of the field will increase under low-speed conditions. The result of this is a dramatic increase of torque at low speeds.

19

Series motors thus find the largest advantage where starting torque requirements are high. Series motors are often used in propulsion applications such as in golf carts, wheelchairs, and electric forklifts. The relatively low cost of these motors, combined with the above characteristics makes them a reasonable choice for mobile robot propulsion.

Compound Motors

A compromise between the characteristics of a series motor, and those of a parallel motor, can be accomplished by placing both series and parallel windings on the stator structure. In most applications, the series winding is much weaker than the parallel winding. A typical speed versus torque curve for a compound motor is shown in Fig. 1-6. Although various curved shapes may be optimal for different applications, a linear speed-torque relationship is generally desirable from the standpoint of servo-control applications.

Commutated Permanent-Magnet Motors

In the past, any motor that was designed to deliver a significant amount of power required an electrical field magnet. Conventional ferrous magnetic materials could not provide a sufficiently intense magnetic field. It is also required that such materials not lose their magnetic characteristics due to vibration, heat or exposure to other fields. This means that a material with a high-flux density, a very low permeability, and a high coercive force is required. Three basic types of magnetic materials have been developed that largely filled these requirements: Ceramic (ferrite), alnico (an aluminum, nickel, cobalt alloy), and rare-earth (samarium-cobalt). The relative merits of these magnetic materials are given in the following table:[2]

Characteristics of Magnetic Materials

Magnetic Material	Relative Magnetic Energy	Density	Notes
Ceramic	1.0	low	Inexpensive
Alnico	3.6	medium	Widely used; reasonable
Rare-Earth	5.1	high	High performance/cost; Very high coercive force

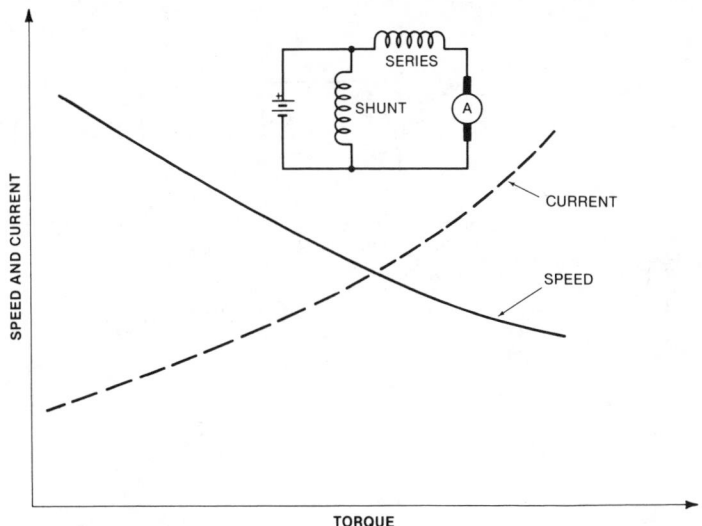

Fig. 1-6. Speed-torque curve for a typical compound motor.

The use of a permanent magnet (for the field) can improve a commutated motor in several ways. First, it modestly improves the efficiency of the motor by eliminating the need to power the field. Additionally, if a low-permeability material is used to produce the magnet, certain undesirable interactions between the armature and field can be eliminated.[1] Because of these advantages, permanent-magnet motors have come to dominate the servomotor market.

The speed-torque curve for a typical ceramic-magnet motor is shown in Fig. 1-7. The *stall torque* (the torque produced by the motor at zero speed) of a permanent-magnet motor is generally higher than the stall torque of a comparable compound motor, but somewhat less than that of a comparable series motor. Notice that the curve is very linear.

As mentioned earlier, brush-type permanent-magnet motors that utilize inside-out construction have recently gained substantial acceptance. These motors are already found in several of the light- and medium-duty industrial robots. The only disadvantage of these motors is that they require two sets of brushes. This is one of the factors that lead to the development of brushless motors (to be discussed later).

Ceramic, alnico, and rare-earth permanent-magnet dc motors of

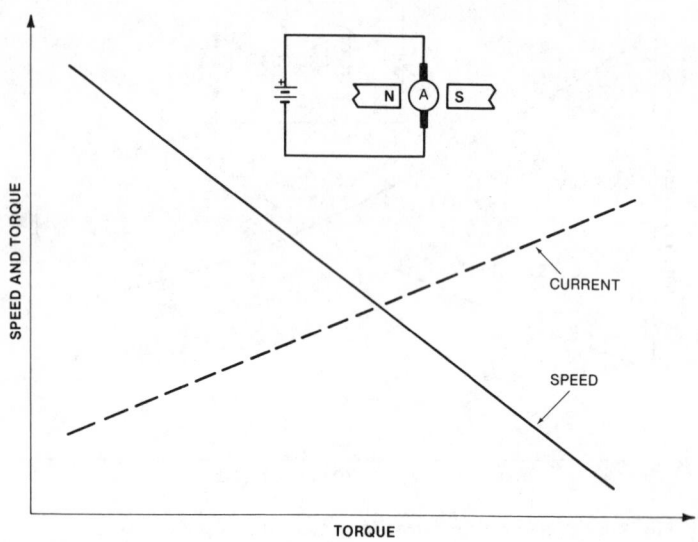

Fig. 1-7. Speed-torque curve for a typical ceramic magnet servomotor.

both conventional and inside-out construction are finding wide usage in light- and medium-duty pick-and-place robots (Fig. 1-8).[M1, M2] With these motors available in powers of 10 hp and up, they are even pushing into the heavier robot markets.

Low-Inertia DC Motors

The inertia of a motor is critical for applications that require quick changes of speed or direction. In high-power systems, the usual method of reducing inertia is simply to make the rotor longer and smaller in diameter. Since the inertia of the rotor is proportional to the fourth power of the radius and only to the first power of the length, this technique can make a very significant impact on the rotor inertia. However, practial limits to the acceleration can be accomplished by this method. The trade-offs faced in the design of motor systems that are located on the moving elements of robot arms often result in the selection of such motors. See Fig. 1-8.

For applications where the power-to-weight ratio of a motor is not critical, extremely fast response motors are available. The several types of these motors are known collectively as moving-coil motors since they have no iron core in the armature. Traditionally these motors have found application in scanning mechanisms, incremental tape drives, disk drives, and in other relatively light-duty uses where response time is critical. Recently heavier-duty versions of printed motors have been incorporated into the design of several light- and medium-duty robotic arms. In these applications, they are generally mounted where the inertia caused by the motor's weight is not a critical problem (i.e., near the base of the robot). Because of their ability to make rapid-step movements and to stop quickly without overshooting, these motors are likely to be used in an increasing number of robotic applications in the future.

Some of these motors are so fast that instead of rating their

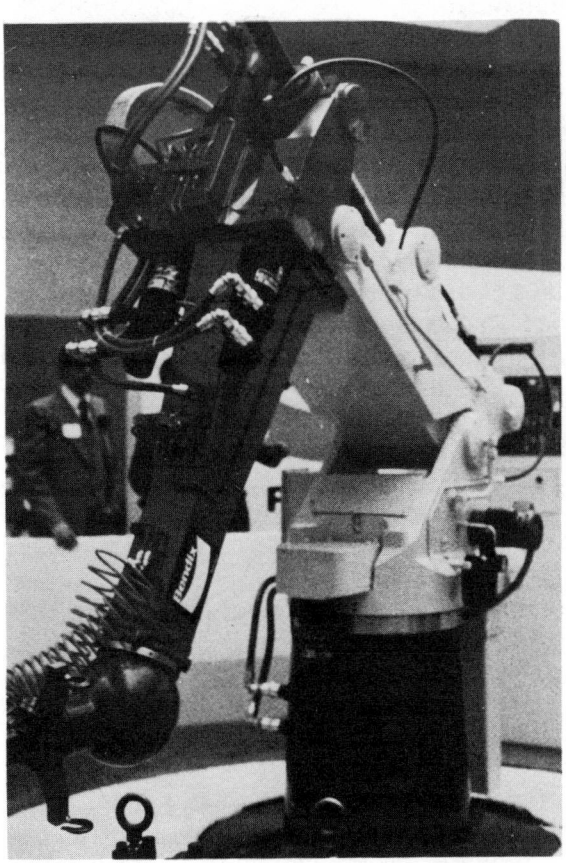

Fig. 1-8. Typical application of rare-earth motors in a pick-n-place robot.

response in radians per second squared of acceleration, the manufacturer may rate them in hertz (Hz) or even kilohertz (kHz). The response of these motors is usually plotted with the output of a tachometer as one axis (in units of decibels) and the frequency of the drive signal as the other axis. This is the same measurement technique used to rate acoustical transducers such as high-fidelity speakers. Motors are available with frequency responses that approach 10 kHz.[M1] Because of the extremely low inertia construction, both the electrical and mechanical time constants of these motors are important.

Printed Motors

One type of moving-coil motor uses a flat, disk-shaped armature (Fig. 1-9).[M3] Earlier versions of this motor used a photo-etching method to print the conductors on the armature, and the name has stuck. Most printed motors are fabricated using lamination techniques. In the printed motor (sometimes called a pancake motor) the armature is connected to a commutator in very much the same way as in a conventional motor.

The field of this type of motor contains many separate magnet pairs, half of which are oriented in one polarity and half of which are aligned in the opposite polarity. The conductors are arranged so that

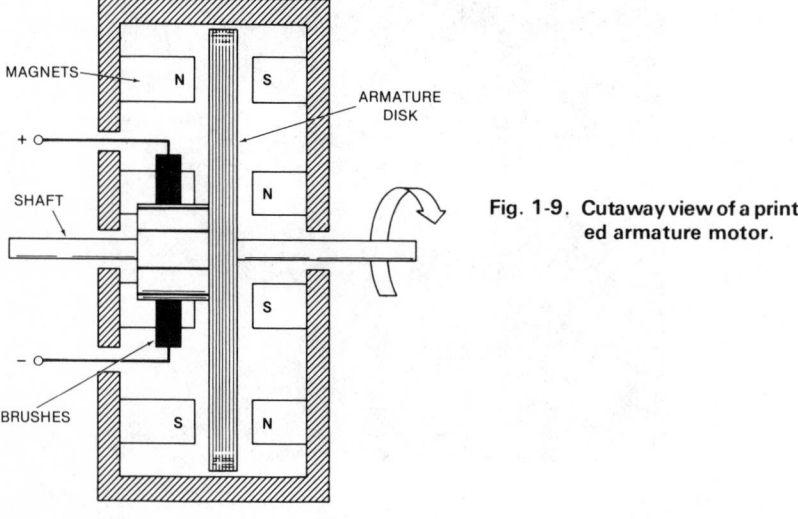

Fig. 1-9. Cutaway view of a printed armature motor.

Courtesy Inland Motors, Industrial Drives Division, Kollmorgen Corp.

current flows out one radial path between a magnet set with one orientation, and returns to the center of the armature on a conductor path that goes between magnets of the opposite orientation. Printed motors are commonly available with torques up to 100 ounce-inches and more and are moderately priced.

Shell and Cup Armature Motors

These low inertia motors avoid two of the inertia contributions of the printed motors:[M1] The relative high inertia of the disk shape and the inertia of the conductors that run about the edge of the disk between the outward and inward conduction paths. By using epoxies and other bonding techniques to form the armature as a cup (Fig. 1-10), the inertia and performance can be made very high. Unfortu-

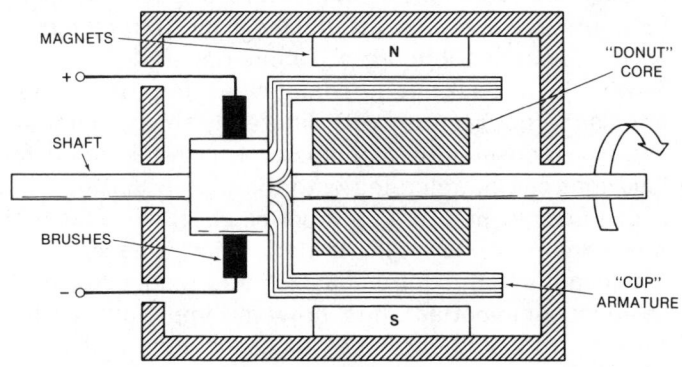

Fig. 1-10. Cutaway view of a cup armature motor.

nately, these fabrication techniques tend to increase the cost of these motors. As with conventional motors, the diameter of the cup has a strong effect on the inertia, and it is generally reduced to the point where other factors limit it.

Servomotor Specifications

A great deal of care must be taken in specifying servomotors because little standardization between manufacturers in the industry occurs. What appear to be the same parameters from two different

25

manufacturers may in fact have quite different assumptions behind them. Some of the more important parameters are given here, along with a brief explanation of their meaning:
(Note: Smaller motors are rated in ounce-inches instead of pound-feet.)

- **Kb (volts/1000 rpm):** This is the back emf constant of the motor, and gives the relationship between speed and the back emf generated.
- **Kt (lb-ft/amp):** Called the *torque constant*, this parameter is the relationship between the current of the motor and the torque it will generate. It should be (but isn't always) taken at the full-load torque and speed of the motor.
- **Tl (lb-ft):** This is the load torque that the motor should be able to deliver continuously at rated speed without overheating.
- **Ts (lb-ft):** This is the stall torque of the motor or the amount of torque that it can deliver at or near zero speed. Some manufacturers rate this value at between 5 and 10 rpm to take into account heat distribution and torque ripple.[M2]
- **Tp (lb-ft):** The peak torque (sometimes referred to as the peak impulse torque) is rated very differently among manufacturers. More conservative manufacturers provide a figure for the load that can be tolerated with the motor starting at rated speed and temperature for a period of ten seconds without any permanent change in the motor's performance. Except in applications that require impulse performance, this value is relatively unimportant. It is, however, important not to confuse it with the other torque ratings just mentioned.
- **Jm (lb-ft/sec-squared):** This is the motor inertia. For maximum efficiency during acceleration and deceleration, the load torque should be matched to this value. When gear reducers and other motion translation techniques are employed, the reflected inertia of the load will be affected. A straight-gear reduction, for example, changes the reflected inertia by the second power of the gear ratio. Several helpful equations are included in Appendix A for the reader who has a deeper interest in this problem.

The speed-torque curve of a motor is probably the most important single bit of information available to the purchaser. When using these curves, the user should check for the conditions under which the curve data was taken. Motors rated for continuous duty should be provided with curves taken at rated temperature. If this is not the case, the purchaser will have to calculate the derating factor. If there is a question of whether the cycles of operation required by the motor constitute continuous duty, then the *thermal time constant* of the motor must be taken into account. If the rate at which the motor is turned on and off is much faster than the thermal time constant, then the power dissipated in the motor can be assumed to be reduced by the ratio of the *on part of the cycle* to the *whole cycle.*

Stepping Motors

With brush-type dc motors, it is necessary to provide some means of monitoring the performance of the motor in order to effect accurate control. For systems where velocity control is desired, this may require a tachometer, while systems requiring position control usually include a position encoder of one type or another. The output of these sensors is fed back to the servo amplifier or computer to *close the control loop.* More will be said about these techniques later in this chapter. For now, it is sufficient to note that they are required because only a rough correlation between the power input to the motor and the velocity at which it moves exists. Normal heating of the motor alone will typically cause a variation in performance of 40% or more.

Because of the complexities associated with feedback systems, the concept of *open-loop control* is very appealing. The stepper motor offers such a possibility although it is not without its own disadvantages. Stepper motors (Fig. 1-11) [M4, M13] consist of a permanent-magnet rotor with either two, three, or four sets of coils called *phases* placed around the rotor in a manner similar to the inside-out construction mentioned earlier (see Fig. 1-2). Instead of these phases being connected to brushes, they are driven by external logic. This drive logic does not normally receive any information back from the motor, but instead the phases are driven sequentially and it is assumed that the motor responds accordingly.

The number of windings in each set of coils and the number of magnets around the rotor determine the *step angle.* The three basic techniques for driving a stepper motor are the full-step mode, the half-step mode, and the microstep mode. In the full-step mode, one phase at a time is turned on in the sequence that will cause the desired rotational motion. In the half-step mode, an intermediate step is generated in between every full step by energizing the next phase

Fig. 1-11. Light-duty stepping motors.

before removing power from the previous phase. The result is to generate a magnetic vector that lies halfway between those of the two adjoining phases. This vector is stronger than a single-phase vector, but less than twice as powerful. For this reason, the motor will exhibit increased torque in the half-step mode, but it will have a slightly reduced efficiency. The microstep mode generates a sequence of vectors between adjacent steps and will be discussed in more depth later in this section.

The drive signals required for the three basic types of stepper motors are shown in Fig. 1-12. Notice that the two-phase stepper requires that the polarity of the drive signal across the phases be reversed in order to generate a four-step sequence. This requirement means that eight switching elements are required in the drive circuit for a two-phase motor, while only three are required for a three-phase stepper, and only four are required for a four-phase stepper motor. For this reason, two-phase steppers are normally utilized in very light-duty applications, where the switching elements are not expensive. In fact, for two-phase stepper motors that require less than 500 mA per phase, a single-chip driver is available at a very low cost.[M5]

Most higher-power stepper motors are of the three- or four-phase types. Typically, these motors are available in step angles of 1.8°, 7.5°, and 15°. These values refer to the full-step magnitude. When these motors are operated in the half-step mode, the step angles will of course be half as large.

As already mentioned, the advantages of stepper motors come with some very significant disadvantages. At low speeds, the motion of the stepper motor is very rough as it quickly moves from one position to the next. The effect of this abrupt motion is for the rotor to oscillate about the new position after each step. If the step frequency corresponds to one of the backward peaks of this oscillation, the motor may occasionally jump back a phase instead of forward. This effect is referred to as *resonance*, and it can completely disrupt the operation of the system. Several techniques may reduce this effect although some of these may not be applicable to certain situations. To reduce resonance in a stepper motor:

1. Operate the motor in the half-step mode.
2. If possible, never operate the motor at a rate in the resonant area.
3. Use frictional or viscous damping.
4. Increase the load inertia to lower the resonant frequency.
5. Use a motor with a smaller step angle.
6. Use a microstepper drive. (This reduces torque slightly.)
7. Use damping resistors between phases. (This reduces efficiency.)
8. Use damping capacitors across each phase. (This use is speed sensitive.)
9. Use intelligent electronic damping (retrotorque damping).[M13]
10. Use a shaft encoder as a feedback mechanism to the drive.

As discussed earlier, all motor specifications should be read carefully, and this is especially true with stepper motors. The label on a stepper motor might, for example, read

Step angle: 15°
Dc volts: 12
Steps per second: 450
Torque: 45 ounce-inches

The unwary purchaser might calculate that the specified step rate would yield a speed of 1125 rpm, and assume that the motor would provide 45 ounce-inches at this speed when powered by a 12-volt battery. This can be an embarrassing assumption! In fact,

29

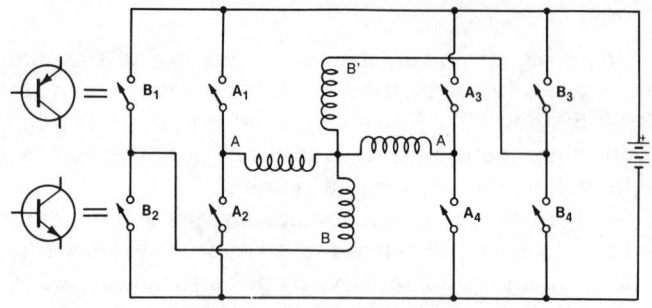

STEP	1.0	2.0	3.0	4.0	1.0	1.0	1.5	2.0	2.5	3.0	3.5	4.0	4.5	1.0
A₁	X				X	X	X						X	X
A₂			X						X	X	X			
A₃			X						X	X	X			
A₄	X				X	X	X						X	X
B₁		X				X	X	X						
B₂				X						X	X	X		
B₃				X						X	X	X		
B₄		X				X	X	X						

FULL-STEP MODE | HALF-STEP MODE

X = SWITCH CLOSED

(A) Driving the two-phase stepping motor.

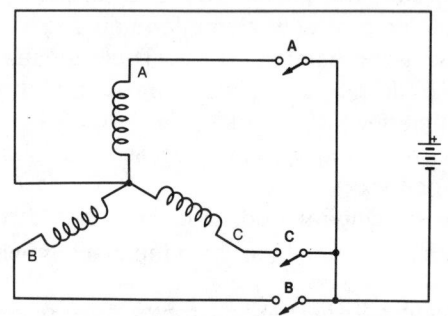

STEP	1.0	2.0	3.0	1.0	1.0	1.5	2.0	2.5	3.0	3.5	1.0
A	X			X	X	X				X	X
B		X				X	X	X			
C			X					X	X	X	

FULL-STEP MODE | HALF-STEP MODE

(B) Driving the three-phase stepping motor.

Fig. 1-12. Drive signals required for the

30

with no load at all, it may not be possible to achieve 450 steps per second with the driver connected to a 12-volt power source. The torque figure for these motors is usually taken at speeds just below or above the resonance zone of the motor (see Fig. 1-14). Worse yet, the torque figure appearing on the label may be the *holding torque.* The holding torque of a stepper motor is the amount of torque that must be applied to the shaft to cause it to rotate when the motor is at rest with one phase energized. A stepper motor with a 45 ounce-inch holding torque might exhibit only one-third of that torque when stepping.

The rated speed, on the other hand, may be the top speed of the motor (i.e., the speed at which the motor can sustain no load torque). Even worse, it may be the top speed attainable when the driver is powered by a much higher voltage than the rated voltage. When this is done, a resistor is placed in series with the windings (Fig. 1-13). The use of this higher voltage helps counteract the back emf generated in

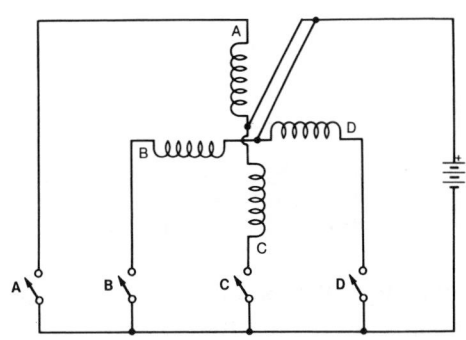

STEP	1.0	2.0	3.0	4.0	1.0	1.0	1.5	2.0	2.5	3.0	3.5	4.0	4.5	1.0
A	X				X	X	X						X	X
B		X					X	X	X					
C			X						X	X	X			
D				X							X	X	X	
	FULL-STEP MODE					HALF-STEP MODE								

(C) Driving the four-phase stepping motor.

three basic types of stepper motors.

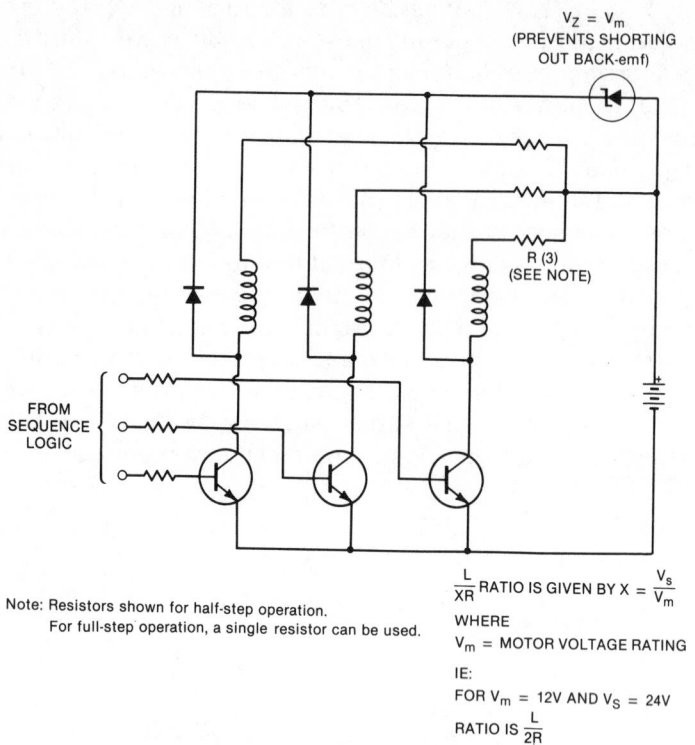

$V_Z = V_m$
(PREVENTS SHORTING
OUT BACK-emf)

R (3)
(SEE NOTE)

FROM
SEQUENCE
LOGIC

Note: Resistors shown for half-step operation.
For full-step operation, a single resistor can be used.

$\dfrac{L}{XR}$ RATIO IS GIVEN BY $X = \dfrac{V_s}{V_m}$

WHERE

V_m = MOTOR VOLTAGE RATING

IE:

FOR V_m = 12V AND V_S = 24V

RATIO IS $\dfrac{L}{2R}$

Fig. 1-13. Practical driver circuit with efficient snap-back protection. (Note: Resistors are shown for half-step operation. For full-step operation, a single resistor can be used.)

the windings as the rotor steps. The ratio of the series resistor to the motor impedance is referred to as the L/R ratio. Stepper motors are sometimes specified with values of up to L/8R. If this were the case with the motor in the earlier example, it would require a 96-volt supply just to achieve the rated speed. To make high L/R ratios more attainable, many steppers are designed with voltage ratings as low as 2 volts. Such motors may actually be intended for operation from 12- or 24-volt supplies.

To help prevent designers from making these mistakes, at least one manufacturer[M4] has published a designer's guide for stepper motors.[3] The only way to safely select any motor, and especially a stepper motor, is to use the specification data in conjunction with the torque-speed curves. With stepper motors, these curves will vary widely for any given motor as shown in Fig. 1-14, depending on the type of drive used to compile the data. Unfortunately, the technique of using high L/R ratios is very inefficient, and the resistors are bulky.

Several sophisticated driving techniques[M7] can greatly enhance the performance of a stepper motor besides the use of high L/R ratios. One of the techniques for enhancing performance is to divide each step pulse into a large number of short, high-voltage subpulses. In between these subpulses, the current through the snap-back diodes is monitored.[M4] When the current falls below a certain level, the controller generates another subpulse, and the process is repeated. This technique avoids the inefficiency of the series-resistor technique and provides a higher level of motor performance. Unfortunately, this type of controller is more complex and expensive than the series-resistor type.

Systems that utilize stepper motors usually do not contain elaborate tachometers or position encoders. Instead, it is assumed that the motor responds properly to each step. The velocity is known (because it is commanded), and position can be determined by counting the

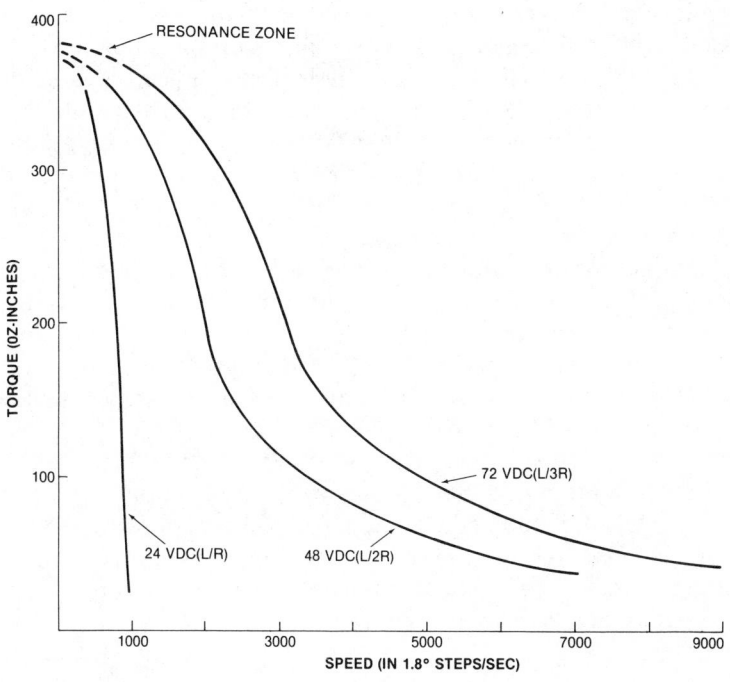

Fig. 1-14. Stepping motor speed-torque curves for various L/R ratios.

33

number of steps taken in each direction from a known reference. Generally, this reference is a limit switch or a mechanical stop. When the system is initialized, the motor is run in the direction of the reference until the switch is activated, or until enough steps have been taken to assure that the system is stalled against the stop. The system can then calculate the position of the mechanism, *as long as the motor does not slip phase.*

The position resolution and smoothness of a stepper motor can be improved by the use of microstepping. This technique divides each step into many microsteps. Each microstep changes the ratio of the drive currents between adjacent phases by a very small amount. Like other enhancement techniques, microstepping offsets some of the economy that was gained by using a stepper motor instead of a servomotor.

Some manufacturers specialize in stepper motors with very small step angles (less than 1°), [M7] and in controllers that further enhance the resolution. These systems are very useful in the automated manufacture and testing of very small items.

In addition to the other precautions that must be observed when stepping motors are employed, a stepper-motor system must be accelerated and decelerated at a rate that allows the motor to overcome the system inertia. If a stepper motor is dynamically overloaded, it will slip phase. At least one manufacturer offers an intelligent four-phase stepper-motor controller in a single LSI integrated circuit.[M6] This device relieves the main processor of a system from the routine tasks associated with velocity ramping and phase commutation.

Several manufacturers of ultralight-duty robot arms use stepping motors to reduce both the cost and programming complexity of their systems. Using stepping motors is a natural decision given the design goals of such robots, but the user must take care never to overload these robots. A robot driven by stepper motors will become very erratic both when phase slipping occurs, and afterwards, as all position references become offset by the amount of the slippage.

Stepper motors find their most suitable application where loads are well below the motor capabilities. The ability of the stepper motor to move at extremely precise speeds, and at a wide range of speeds (assuming resonance is avoided) makes them popular for many types of positioning applications.

Since phase slippage is such a significant problem with stepper motors, it is tempting to equip the motor with a shaft position encoder and to feed this signal back to the controller. In this way, resonance problems could be reduced, and phase slippage could be eliminated. This is done by at least one manufacturer, but in general the applications that require this type of measure are those where reasonable high torque is needed.[M13] In these applications, a brushless motor might be the best choice.

Brushless DC Motors

Brushless dc motors (Fig. 1-15) can be thought of as a cross between an inside-out motor and a stepper motor with a shaft encoder.[M8] Most of these motors are three-phase motors. Those familiar with synchronous ac motors will recognize the concept of driving three motor phases with separate sine-wave voltages (120° apart in phase) in order to generate a rotating magnetic vector. With synchronous motors, the armature becomes synchronized to this vector when running speed is achieved. Also, a similarity exists between a three-phase stepper motor driven in the half-step mode and the brushless dc motor.

There are subtle but very important differences between the brushless motor and these other motors, however. When a synchronous ac motor is accelerating to running speed, it is slipping phase because the frequency of the power source is fixed (i.e., 60 Hz). This slippage badly reduces the torque available from the motor during acceleration. A synchronous motor must run at *line speed* to develop full torque. On the other hand, a stepper motor normally accelerates in phase with its drive signals, but the driver does not usually take into account the response of the motor to its commands.

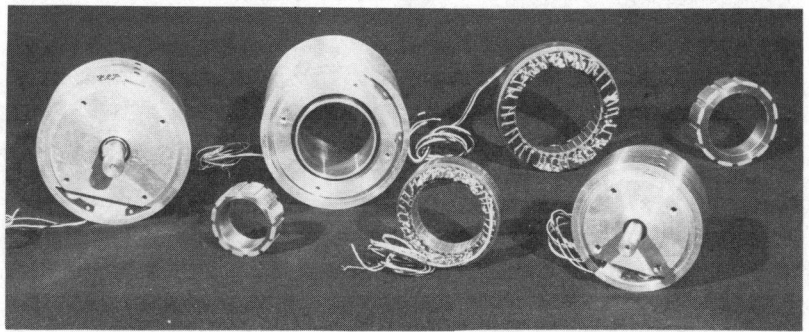

Courtesy International Scientific Industries.

Fig. 1-15. Construction of a brushless rare-earth dc motor.

Brushless dc motors use a shaft encoder to close the loop between the controller and the motor. The inside-out motor with Hall-effect brushes (discussed earlier) is in fact a simple type of brushless motor. The higher performance brushless motors, however, make use of much higher resolution information about the shaft angle of the motor. By doing this, the driver is capable of generating smooth, highly efficient sinusoidal drive signals. This process is similar to the microstepping process discussed earlier, except that by using position feedback, the controller can optimize the drive signal in such a way as to greatly increase the performance and efficiency of the motor.

Unlike any other type of system, the driver for a brushless dc motor can be made to take into account the speed and load of the system, and it can adjust both the waveshape and phase of the drive signals. This is a very powerful combination.

When running at low speeds, brush-type dc motors (Fig. 1-16) have a tendency to *cog*. This effect (also referred to as *torque ripple*) can be very undesirable where precise control is required. The more the inertia is reduced in such systems (to increase performance), the worse the torque ripple becomes. When driven from a high-resolution controller, a brushless dc motor can be made to have very low torque

Fig. 1-16. Construction of a brush-type rare-earth motor.

ripple. For all of these reasons, brushless dc motors are receiving a lot of attention by robot manufacturers. Unfortunately, drivers for these motors are not widely available, and their cost is correspondingly high. This situation should improve in time, but because of their relative complexity, the controllers for these motors will always be relatively expensive.

TACHOMETERS AND POSITION ENCODERS

As already discussed, in order to obtain accurate control over the performance of a motor, it is necessary to feed back a signal to the controller that is proportional to the factor (speed or position) that is to be controlled. This feedback will allow the controller to compensate for such factors as changing loads, temperature, and supply voltages.

Analog Tachometers

An analog tachometer is similar to a permanent magnet dc motor running as a generator. Since, however, there is no requirement to produce power as a generator, certain trade-offs are made in the construction of these devices. These differences provide a generator output that is very accurately proportional to the speed of the shaft. Some motors are constructed with both the regular armature windings and the tachometer windings in the slots of the rotor.[M1] The armature and tachometer commutators of such motors are generally at opposite ends of the rotor. More often, the tachometer generator is housed in a separate case that can easily be piggybacked onto the mating motors.

The outputs of such tachometers have inherent ripple at the commutation rate (Fig. 1-17), and must therefore be passed through a low-pass filter before being read by any circuit that does not contain inherent input integration. A computer, for example, that was to read the speed of a motor through a fast analog-to-digital (a/d) converter, would require such a filter to prevent erratic readings.

The ripple of these tachometers limits their low-speed performance. If a very low frequency filter is used on an analog tachometer (to extend its low-speed usefulness), the added feedback delay might cause control instability at higher speeds.

In addition to the ripple problem, analog tachometers suffer from another shortcoming. Although the output of an analog tachometer

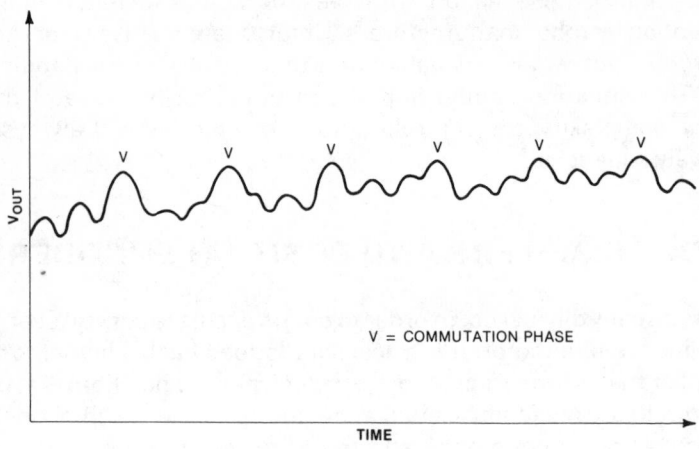

Fig. 1-17. Typical ripple output of an analog tachometer.

can be integrated to give a rough position indication, it is not accurate enough for this purpose in applications such as robotic systems. What is needed in such applications is a tachometer that is digital in nature. Such a tachometer would not exhibit an accumulating error when it was used to determine distance.

Pulse Tachometers

With the increased use of microprocessors and other digital systems in machine control applications, pulse tachometers are rapidly gaining popularity. The simplest form of pulse tachometer is the single-channel type of output, which is shown in Fig. 1-18A. Light is transmitted through a transparent disk printed with uniformly spaced opaque radial stripes. As the disk rotates, either the rate of the pulses or the time between pulses may be measured to determine the speed of the shaft. Since the circuit to which the tachometer is attached cannot determine the direction of the motion, this type of tachometer is only useful for position calculations in systems that operate in a single direction.

By adding a second channel to the pulse tachometer just as described (see Fig. 1-18B), the direction corresponding to each pulse can be determined. The two channels are said to be in *quadrature* (90° apart in phase). When the shaft is rotated in one direction, the signal from channel A will make positive transitions while the channel B signal is positive (logically true). When the shaft is rotated in the opposite direction, the positive transitions of channel A will occur while the channel B input is in the low (logical zero) state. The circuitry required for translating these signals into a usable form is referred to as a *pulse tach decoder*, and it can be implemented in either hardware or software. When this function is done in software, one or both channels are made to interrupt the processor. At high speeds, this can overload the processor and should thus be avoided.

A third channel is sometimes available on a pulse tachometer. This channel is called the index, and it provides a single pulse per revolution of the shaft. When the system is seeking a reference position, the index provides an absolute reference as to the shaft position. The use of this signal can greatly improve the resolution of the reference.

Pulse tachometers not only allow the measurement of speed and relative distances, they are also usable at much wider ranges of speed than analog tachometers are. This low-speed performance is made possible by the use of pulse-period measurements. When stable control at very low motor speeds is required, a tachometer with

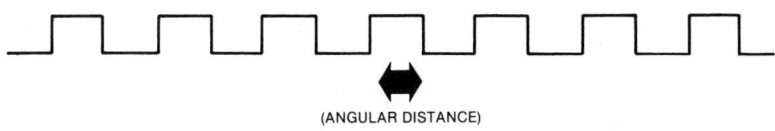

(ANGULAR DISTANCE)

(A) Output of a single-channel pulse tachometer.

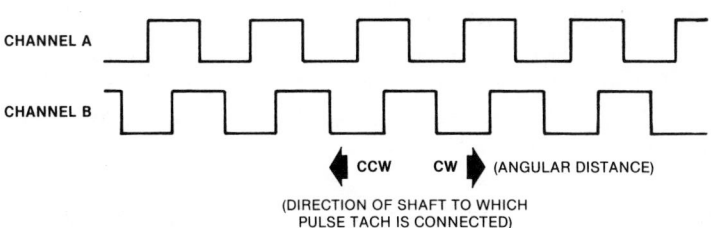

CHANNEL A

CHANNEL B

CCW   CW   (ANGULAR DISTANCE)

(DIRECTION OF SHAFT TO WHICH
PULSE TACH IS CONNECTED)

(B) Output of a dual-channel pulse tachometer.

Fig. 1-18. Two types of pulse tachometers.

more pulses per revolution must be used, or the connection between the tachometer and the motor must be geared up.

The mechanical simplicity of pulse tachometers should make them less expensive than analog tachometers, but this is not necessarily the case. The difference in the ages of these products, and the difference in the levels of competition between manufacturers has kept pulse tachometers relatively expensive. As the market for these devices grows, the price should drop.

Position Encoders

As with tachometers, there are both analog and digital techniques for making absolute angular and longitudinal distance measurements. Although there is a tendency toward digital techniques, hazards in their use are not always foreseen. These problems will be discussed more in the section on motor control.

Potentiometers

Popular in the past because of their simplicity, potentiometers are seldom found in modern systems except where infrequent operation is expected. Although relatively reliable, high-quality, wire-wound potentiometers are available for this kind of use, the inherent frictional wear of these devices makes them noisy and less reliable than other techniques.

Resolvers and Inductosyns

Resolvers are essentially transformers with rotating secondaries. The primary winding of the resolver is excited by an ac voltage. Although resolvers have traditionally been designed for excitation at power-line frequencies (usually 60 or 400 Hz), robotic applications usually require higher excitation frequencies. Excitation frequencies in the range of 5 kHz are not uncommon.[M9]

The output of the secondary of the resolver is proportional in amplitude to the sine of the shaft angle. Negative values (180° to 360°) are represented by a 180° electrical phase shift from the primary signal. The rotational speed of the shaft must be much lower than the frequency of the electrical excitation if the output is to be meaningful (thus the high-excitation frequencies already discussed). Resolvers exhibit good accuracy and are moderate in cost. When they are used to measure robot joint angles, resolvers may be geared up to increase the measurement resolution. Resolvers are also available with more than one pole. A resolver with two poles would provide the same output signal as a single-pole resolver that was geared up by a factor of two.

Inductosyns are similar to resolvers, except the phase of the output is proportional to the shaft angle. These devices are more accurate and expensive than resolvers.

Laser Interferometers

Laser interferometer devices use the interference patterns produced by recombining two beams from a single laser. By counting the peaks of the interference pattern, very accurate longitudinal distance measurements can be made. The technique is very expensive and is thus used only where the accuracy of other systems is inadequate.

Shaft Encoders

These devices are fairly popular for making absolute angular distance measurements in robotic applications, especially when the control loop is closed through a computer. Similar in construction to the pulse tachometer, they contain many more channels. The output of the channels is a digital representation of the shaft angle.

One of the problems associated with such encoders occurs when the shaft angle is such that the detectors are lined up with the very edge between two readings. If a conventional code is used (such as binary or binary-coded decimal), gross errors can occur as a result of some of the channels reading the stripe values from one side of the transition and others reading the values from the other side of the transition. This problem can be corrected by using a code such as the Gray code (Fig. 1-19). This code is designed so that only a single-bit change from one code number to the next occurs.

Some shaft encoders contain internal mechanical counting mechanisms that keep track of the number of revolutions of the disk. These devices allow the exact longitudinal position of the device that the motor is driving to be determined without the need for references.

41

DECIMAL	BCD				EXCESS 3 GRAY			
	D	C	B	A	D	C	B	A
0	0	0	0	0	0	0	1	0
1	0	0	0	1	0	1	1	0
2	0	0	1	0	0	1	1	1
3	0	0	1	1	0	1	0	1
4	0	1	0	0	0	1	0	0
5	0	1	0	1	1	1	0	0
6	0	1	1	0	1	1	0	1
7	0	1	1	1	1	1	1	1
8	1	0	0	0	1	1	1	0
9	1	0	0	1	1	0	1	0

Fig. 1-19. Comparison of binary-coded decimal (bcd) and excess-three gray codes.

Note: Only one bit changes at a time with the gray code.

Glass Scales

These devices are the linear equivalents of the shaft encoders. Like the shaft encoder they contain many channels of optical sensors that detect a scale painted on a transparent "ruler." Glass scales are also available that have only one or two channels and are the linear equivalent of pulse tachometers. Because of their long size and their vulnerability to damage, these devices are not commonly seen in industrial applications.

Servomotor Control Techniques

Now that the basic components of electrical servo systems have been introduced, we can discuss some of the ways of obtaining precise motion control. The technique that is best for any given system will depend largely on the characteristics required for the application. Since robotic systems most frequently require both velocity and position control, this discussion will concentrate on the techniques for obtaining this kind of control.

Closing the Feedback Loop

The fact that a robot contains one or more computers does not necessarily mean that the servo feedback loop is closed through the computer. In fact, it has been common practice in the past for robot manufacturers to purchase (or construct) analog servo-control assemblies that contain all of the necessary circuitry to provide both position and velocity control of the servo loop. In such systems (Fig. 1-20) the computer provides the servo system with either an analog or digital signal that represents the position to which the motor is to drive the mechanism. If this signal is digital, it is immediately converted to an analog signal by the use of a d/a (digital-to-analog) converter.

When the position input signal of a position controller changes, a position error signal (P_e, Fig. 1-20) is generated at the output of the position summing junction. This signal is applied to an acceleration/deceleration ramp generator. This generator develops a velocity signal that ramps up to maximum velocity and then starts ramping the velocity back down in such a way as to bring the motor to the desired position. For short moves, the motor may not reach maximum speed before the deceleration ramp begins. The actual shape of the ramp may be linear (called trapezoidal), or it may be such that constant horse power or torque is maintained. The approximate shapes of these three curves are shown in Fig. 1-21.[1,2]

The velocity signal generated by the accel/decel circuit is applied to the motor amplifier. The motor amplifier uses a velocity feedback summing junction and a power amplifier to cause the motor speed to agree with the velocity command signal. More will be said about types of power amplifiers later in this discussion.

When both the position and velocity control loops are performed by an analog driver, the processor is usually provided with digital

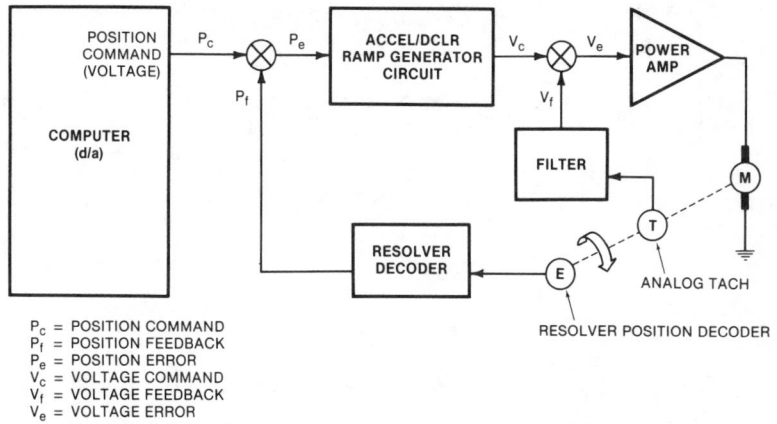

P_c = POSITION COMMAND
P_f = POSITION FEEDBACK
P_e = POSITION ERROR
V_c = VOLTAGE COMMAND
V_f = VOLTAGE FEEDBACK
V_e = VOLTAGE ERROR

Fig. 1-20. An analog position control system.

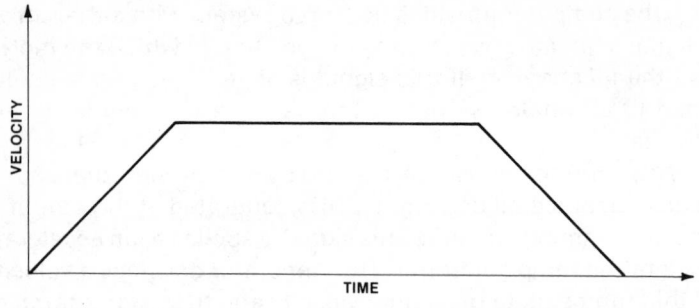

(A) Trapezoidal acceleration/deceleration.

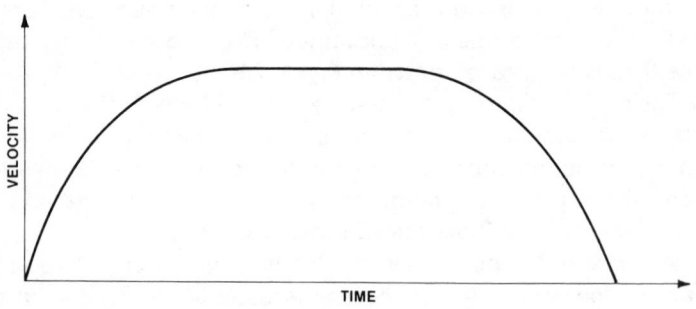

(B) Constant torque acceleration/deceleration.

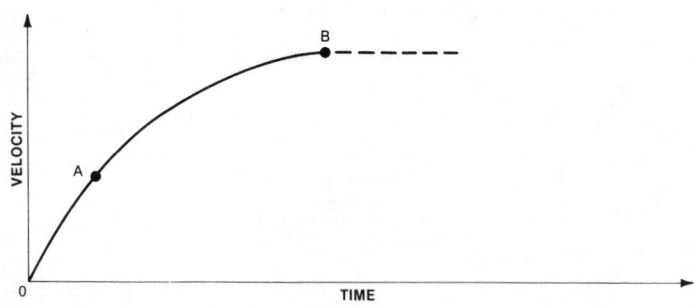

(C) Constant torque from points O to A, constant horsepower from points A to B.

**Fig. 1-21. The shapes of the various signals developed by an acceleration/
deceleration ramp generator.**

signals from the controller called *flags.* These signals indicate such things as the accomplishment of the commanded move and other bits of status information. Additionally, the processor may be connected to a position encoder or tachometer as a means of monitoring the functioning of the driver.

An alternate control technique is shown in Fig. 1-22. In this system, both the velocity and position loops are closed externally with the use of a single-pulse tachometer. The output of the pulse tach is connected to a decoder circuit that generates up and down counting signals to an up/down counter that in turn keeps track of the system position. The computer might also be connected to read (monitor) the position counter.

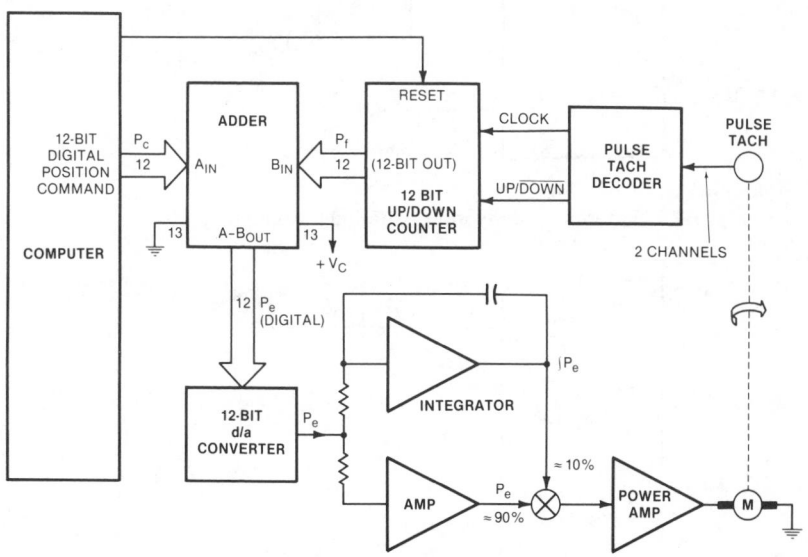

Fig. 1-22. A hybrid position control system.

The digital output of the position counter is subtracted from the digital position command word generated by the computer, giving a digital error value. This value is then converted to an analog form by a d/a converter. The analog error signal is scaled and then applied to the power amplifier. To provide less hysteresis around zero error (where the position error driving signal drops to nearly zero), an error integrator is used. This term is usually small compared to the main error signal.

The adder circuit of Fig. 1-22 could be accomplished in the processor, but since this calculation must be done very frequently, it is not always profitable to do so. All sorts of combinations of hardware and software processing are possible.

As an example of the possibilities available, the position loop or both the position and velocity loops may be closed through the processor (Figs. 1-23 and 1-24). Doing this brings both advantages and

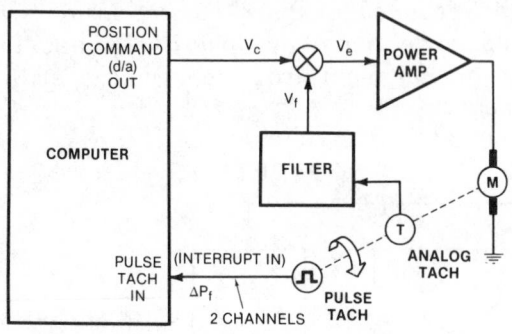

Fig. 1-23. External velocity loop—internal position loop.

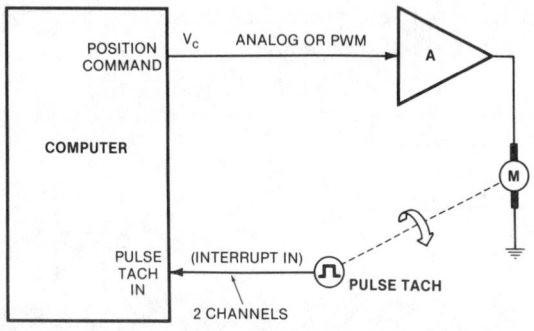

Fig. 1-24. Both velocity and position loops closed internally.

disadvantages, but it is clear that this is the trend for future robots. By performing the closed-loop calculations in the processor, more control flexibility is possible. For example, with adaptive control techniques, the robot can be made to tailor acceleration rates and velocities to changing load conditions, and to take into account the performance of the motor (and the other motors of the manipulator)

46

as heating occurs. In most present-day systems, the acceleration rates and velocities are set at worst-case values, causing a significant loss of efficiency in normal operation. For robots on an assembly line, it may or may not be possible to take advantage of the enhanced capability of adaptive controls because the line speed may depend on the slowest task. The optimization of line performance by synchronization of all of the tasks will be discussed in Chapter 2.

As mentioned earlier, closing control loops through processors also brings disadvantages. One of the biggest problems is referred to as *transport delay*. This refers to the discrete amount of time required for the processor to input the feedback signal(s), calculate the appropriate adjustment, and to output the new command. Furthermore, the digit-control process is not a continuous function, but is instead performed in a sequence of discrete steps. The error contributed by this fact and by the step nature of a/d and d/a conversions is referred to as *quantitization error*. These problems are becoming less significant as microprocessors become faster and less expensive, and as a/d and d/a converters become available in higher resolutions and with faster conversion rates. In fact, microprocessors have become so inexpensive that some robots utilize a separate microprocessor for each motor control.

When computers are used to close the control loop, it is very important that they make their calculations at very precise intervals. If *time distortion* occurs, the system may become erratic. Such control cannot be accomplished using standard high-level languages unless a *real-time executive* program is available for use with the language, or unless some special hardware timer synchronization is used. The programming of such processes is a highly disciplined art.

Servomotor Power Amplifiers

The control of a dc servomotor requires a power amplifier that can supply a variable current to the motor. It should be remembered that the torque of the motor is proportional to the motor current, but the speed is related to the back emf (and thus the applied voltage). Under a constant-load torque, there is a direct relationship between motor voltage and current, but under varying-load conditions the motor current must be adjusted to keep the speed (and thus the back emf) constant.

Three of the most popular basic approaches for implementing dc power amplifiers are shown in Fig. 1-25. The simplest approach (Fig.

1-25A) is a series pass element (shown in this example as an npn power transistor). For clarity, a single-direction drive is shown. Such a circuit uses the output device (a FET or transistor), to control the current through the motor. The fundamental problem with the series pass approach is that a great deal of power is dissipated (thrown away) in the pass element. For any given current, the sum of the voltages across the motor, current sense resistor (not shown), and series pass element must be equal to the supply voltage. This defines

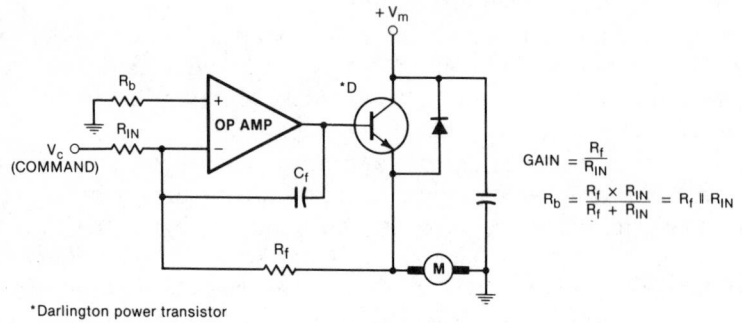

$$GAIN = \frac{R_f}{R_{IN}}$$

$$R_b = \frac{R_f \times R_{IN}}{R_f + R_{IN}} = R_f \parallel R_{IN}$$

*Darlington power transistor

(A) A simple Class-A motor speed control (unidirectional).

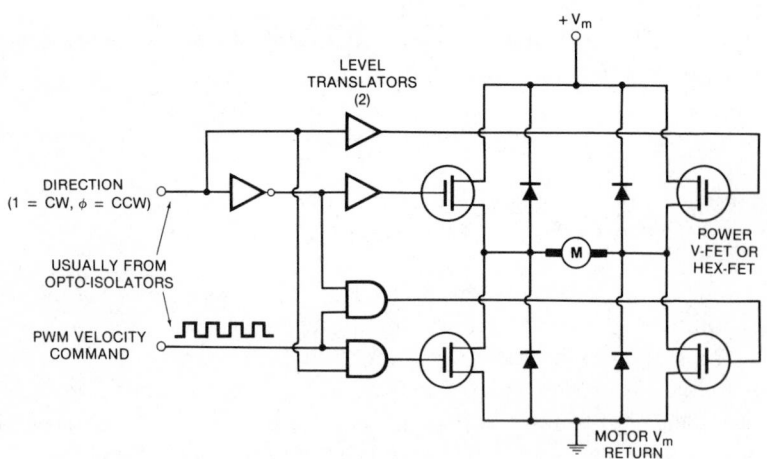

(B) A bidirectional pwm power amplifier using the "H" bridge technique.

Fig. 1-25. The three most popular basic approaches

the voltage drop across the series pass element and the current through it. This power is not only lost, but it is lost in the form of unwanted heat that must be removed from the controller. Because of this problem, the series pass type of power amplifier is seldom used on any motor that requires a significant amount of power.

At intermediate power levels, the pwm (pulse-width modulation) type of amplifier is very popular. These amplifiers (Fig. 1-25B) turn the current through the motor fully on for a portion of a short, fixed time period, and fully off for the remainder of the period. The average current delivered to the motor is controlled by varying the ratio of the on portion to the full period. The switching elements (either power FETs or transistors) are thus always in one of two states: fully on or fully off. The power dissipated in the switching elements is proportional to the product of the current through them and the voltage across them. When an element is off, there is no current through it, so no power is dissipated. When an element is on, there is a lot of

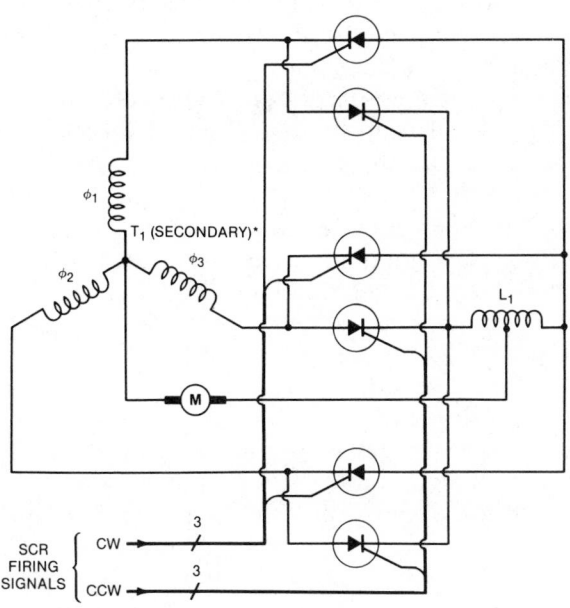

*Primary connected to 3-phase line (not shown).

(C) A high-power, three-phase, SCR power amplifier (simplified).

for implementing dc power amplifiers.

current, but very little voltage drop across it, so the power dissipation is minimal. The efficiency of this type of controller is thus much better than that of the series pass type. The configuration shown is referred to as an "H" circuit. For any given direction of motor travel, one element of the top of the "H" is turned on, and the opposite element on the bottom of the "H" is driven by the variable-duty cycle square wave. The square wave may be generated directly by the computer, or it may be generated by using an analog comparator to compare the velocity command input voltage to a fixed sawtooth reference waveform. Extreme care must be taken that two elements on the same side of the "H" are never energized at the same time especially as the direction command is changing.

In some pwm systems, the current through the motor is never turned off. Instead, the current is reversed for opposite half-cycles of the pwm signal. This causes the motor to exhibit better control, especially around stationary positions and in the face of external load disturbances. Unfortunately, power efficiency and motor heating are both adversely affected by the use of this type of control.

The third type of drive is the SCR (silicon controlled rectifier) (Fig. 1-25C). This drive is very popular in high-power applications. The SCR drive is similar to the pwm drive in that the control elements are operated in a switched mode, and power control is done by duty-cycle manipulation. SCR-type drivers are almost always used with three-phase ac power lines. On each leg of the input line, there are two sets of control SCRs in opposite polarity. To drive the motor in one direction, the SCRs in one orientation are fired (turned on), and for the other direction, the other set of SCRs is fired. The amount of power delivered to the motor is varied by firing the SCRs at various times during the cycle of the line voltage. Once an SCR is fired, it will remain turned on until the end of the cycle (the polarity across the device is negative), or until it is turned off by the SCR of the next phase assuming the load. By firing the SCRs earlier in the cycle, more power is delivered to the servo motor (through the inductor L1). In actual SCR controllers, firing signals are isolated from the controller by the use of pulse transformers or optoisolators (not shown in Fig. 1-25).

In high-power applications, the SCR type of driver tends to be less expensive and more reliable than its pwm counterpart. Unfortunately, there are certain disadvantages to the SCR controller. Perhaps the most significant disadvantage is that the control frequency capability is relatively low as a result of the circuit having to be synchronized to the ac power line. With three-phase circuits, this problem is less (by a factor of three) than with single-phase circuits, so single-phase implementations are seldom used. To further increase the bandwidth of these amplifiers, some manufacturers use special transformers that generate six phases (at 60° apart in phase) from a three-phase input.[M2] These extra phases are connected exactly as are the three phases in the implementation shown, thus doubling the control bandwidth.

Two additional problems with the SCR driver are nonlinearity and rfi. The nonlinearity problem is caused by the sinusoidal relationship between the firing delay time (called the firing angle) and the output. The use of extra phases reduces this problem. The rfi problem is caused by the regenerative effect inside the SCRs as they fire. This effect causes a very sharp increase in the current as the device turns on. Such sudden changes in current generate a large number of noise signals at frequencies that can disrupt radio communications. For this reason, elaborate rfi filtering is usually required.

A Tachless PWM Controller

Using the pwm scheme, it is possible to eliminate the tachometer by measuring the back emf of the motor at the end of the off cycle of the control period (see Fig. 1-26). When the current is first disrupted in the off cycle, a voltage snap-back spike will occur due to the motor inductance. This spike should be allowed to dissipate through the snap-back diode before the measurement of the back emf is taken. By using this technique, very accurate control is possible, provided that the duty cycle is never so close to 100% that there is not enough time to make the back emf measurement.

Also shown in the system of Fig. 1-26 is the use of variable dynamic braking. During braking, the upper two elements (FETs) of the "H" are turned off, and both the bottom elements are pulsed by the variable-duty cycle pwm signal. The reverse diode across one or the other of the bottom elements (depending on the direction of rotation) will conduct in concert with the opposite switch element. The effect of this process is to short out the back emf, thus producing a significant drag on the motor.

HYDRAULIC MOTORS AND CONTROLS

The two most common types of hydraulic actuators are the cylinder and the rotary actuator. Rotary actuators (Fig. 1-27) have

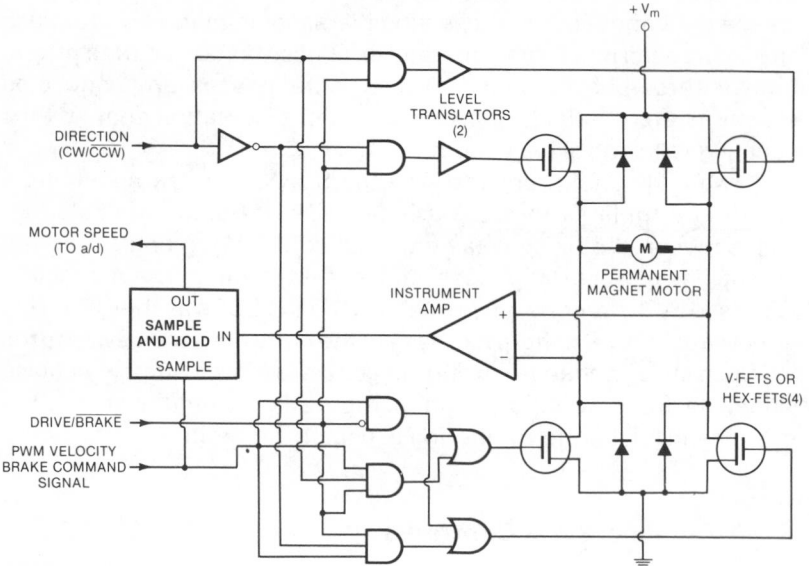

Fig. 1-26. A "tachless" pwm amplified with dynamic variable braking.

come to dominate the hydraulic robot market. These devices are nothing more than a rotary vane separating two ported chambers. As hydraulic fluid is displaced from one chamber to the other the vane rotates, driving the output shaft.

Rotary actuators are rated in radians per cubic inch displacement. The torque delivered by a rotary actuator is proportional to the product of the rating (radians/cubic inch) and the pressure (pounds/square inch). Torque ratings of 1000 foot-pounds are not uncommon in these actuators.[M10]

The control circuit for a hydraulic actuator is similar in structure to that of an electrical motor, except that the power output device is replaced with a variable four-way valve. This valve can be driven by either an analog signal or a pwm signal. When an analog signal is used, there is a tendency for the valve to stick in the presence of very small command changes. To prevent this, manufacturers usually add a slight dither signal to the valve command voltage.[M9] This is usually a sinewave or squarewave signal just above the upper edge of the frequency response of the valve (200 to 400 Hz) and at a level of about 10% of the command signal. The dither signal breaks the friction of the valve, allowing much smoother control.

Because the demand on the hydraulic pump of a robot varies greatly as a robot is operated, special *variable-displacement pumps* have been designed. These pumps have the ability to change their displacement by a small hydraulic servomechanism connected to adjust the angle of a "wobble plate." This allows the pumping of variable volumes of fluid at a constant pressure, according to demand.

AIR MOTORS

Recently, pick-and-place robots that use air motors have become available.[M14] Air motors consist of a rotary vane driven by air pres-

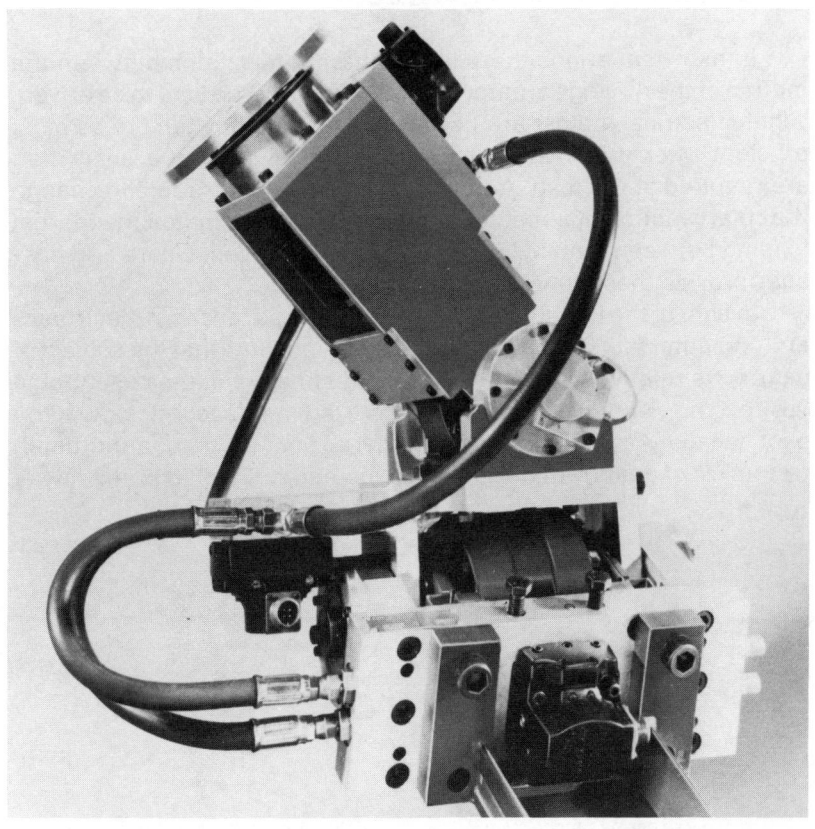

Fig. 1-27. A hydraulic "wrist" actuator.

sure. Because of the compressibility of air, these motors are usually connected to the load through a high-ratio gear reduction unit.

The air supply to the rotary air motor is controlled by a four-way valve, much as in hydraulic systems. The air flow is usually controlled by driving the valve with a pwm signal. The nature of this signal provides a certain amount of inherent *dither*.

Air motors have several advantages, including:

1. inherent cooling,
2. cleaner than hydraulics, and
3. low cost.

GEARS AND LINKAGES

In many situations there is an inherent mismatch between the motion of the driving actuator and the motion of the load to be driven. When electrical motors are used, for example, they tend to be most efficient when driven at higher speeds and lower torque levels than are required at the load. Additionally, it is often desirable to change the rotary motion of a motor to linear motion or to match the inertial loading between a motor and the load to be driven. Gears and linkages provide these functions, but at a cost.

In light of the incredible rate of the recent advances in electronics and computers, the reader may mistakenly assume that the subject of gearing is relatively "old hat." This is distinctly not the case. Some very exciting improvements have occurred in this technology in the past decade. As with other topics, there is not room for an in-depth treatment of gear train design here. Instead, we will look at a few of

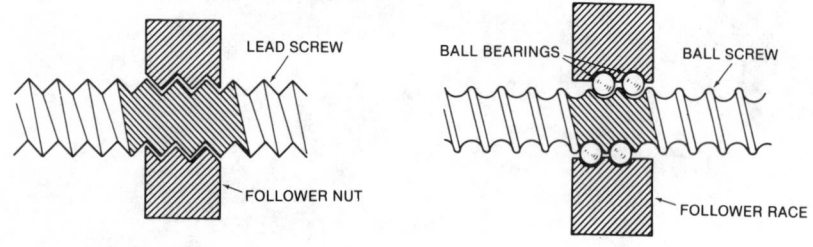

(A) Lead screw mechanism.   (B) Ball screw mechanism.

Fig. 1-28. Lead screws and ball screws.

the more popular and interesting types of gear systems, and discuss their applicability to various robotic applications.

The characteristics of interest in a gear system will vary greatly with the particular application. For a gear assembly that is to be mounted on the end of a manipulator arm, the weight, size, and backlash will be critical. In such an application, the efficiency of the gear assembly may be of secondary importance to these characteristics. On the other hand, efficiency may take priority over these other factors for a gear reduction unit that is to be used in the drive of a mobile robot.

Lead Screws and Ball Screws

Lead screws and ball screws are used to translate rotational motion to linear motion. A lead screw (Fig. 1-28A) is nothing more than a rotating threaded rod with one or more follower nuts. When a lead screw is used, the translation of the load inertia to the motor is inversely proportional to the square of the screw pitch (see Appendix A). Unfortunately, lead screws exhibit relatively high frictional losses. When a tight fit is used to reduce backlash in a lead screw, the frictional losses increase. For this reason, it is normally necessary to heavily grease lead screws, and this is not desirable. Lead screws are usually used where heavy wear is not expected and where cost is crucial.

Ball screw drives (Fig. 1-28B) use a set of ball bearings in a special race instead of a follower nut. The threads of a ball screw drive rod are rounded to fit the ball bearings. The pressure at which the balls ride against the drive shaft can be adjusted to minimize backlash without prohibitively increasing frictional losses. Ball screw drives are much more efficient than are lead screw drives; they do not require heavy lubrication, and they will operate for extended periods without significant wear. Ball screw drives are found in a wide variety of automated systems.

Harmonic Drives

Harmonic gear assemblies (Fig. 1-29) consist of two outer gear splines (bands) and one inner gear spline.[M11] One of the outer splines (the dynamic spline) has the same number of teeth as the inner (flex) spline, and is driven by, and at the same speed as, the inner spline. The output shaft is connected to the dynamic spline.

The second outer spline has two more teeth than the inner spline, and is grounded so that it does not rotate. The input shaft of the harmonic gear assembly drives an eccentric wheel called a *wave generator*. The eccentricity of the wave generator causes a portion of the flex spline to be engaged with the outer splines. This engaged area rotates with the wave generator's action. The engagement of the flex spline with the fixed circular spline causes it (and thus the

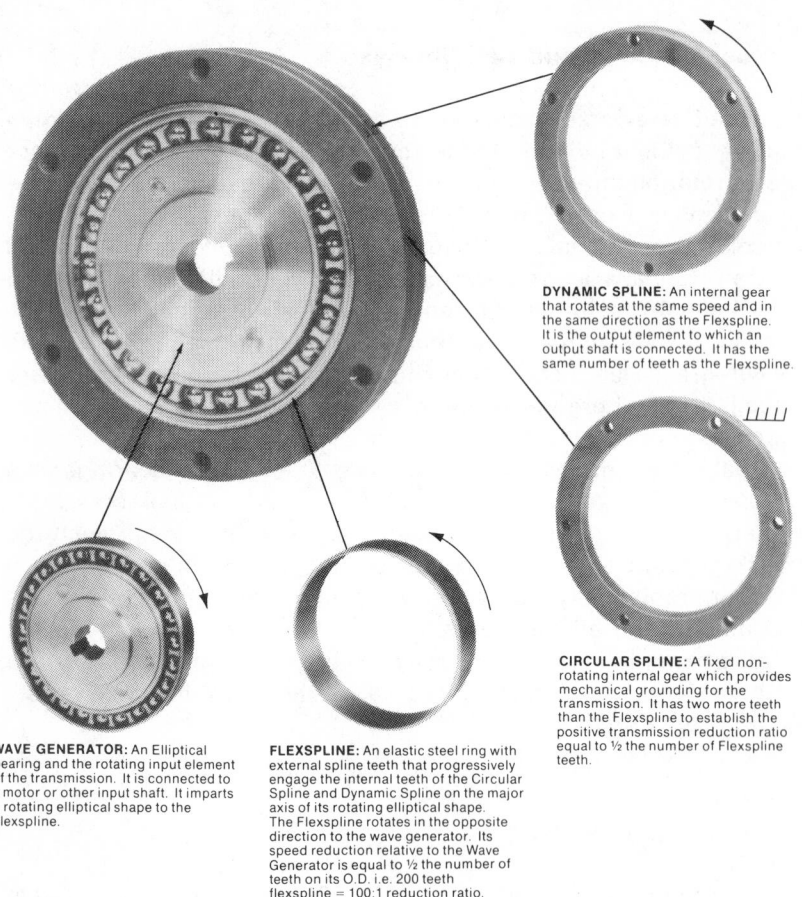

DYNAMIC SPLINE: An internal gear that rotates at the same speed and in the same direction as the Flexspline. It is the output element to which an output shaft is connected. It has the same number of teeth as the Flexspline.

CIRCULAR SPLINE: A fixed non-rotating internal gear which provides mechanical grounding for the transmission. It has two more teeth than the Flexspline to establish the positive transmission reduction ratio equal to ½ the number of Flexspline teeth.

WAVE GENERATOR: An Elliptical bearing and the rotating input element of the transmission. It is connected to a motor or other input shaft. It imparts a rotating elliptical shape to the Flexspline.

FLEXSPLINE: An elastic steel ring with external spline teeth that progressively engage the internal teeth of the Circular Spline and Dynamic Spline on the major axis of its rotating elliptical shape. The Flexspline rotates in the opposite direction to the wave generator. Its speed reduction relative to the Wave Generator is equal to ½ the number of teeth on its O.D. i.e. 200 teeth flexspline = 100:1 reduction ratio.

Courtesy USM Emhart.

Fig. 1-29. Harmonic gear assembly.

dynamic spline and the output shaft) to rotate at a rate that is equal to the input shaft speed multiplied by one-half the number of teeth on the flex spline.

Harmonic drives feature very light weight and small size, and are available with reduction ratios from 80:1 to 160:1. They are a natural match to the pancake printed motors discussed earlier. The efficiency of these drives is typically 50% to 65%, which is not particularly good. The backlash of these assemblies goes from 8 to 36 arc-minutes depending on the reduction ratio and the size. The heavier units offer less backlash, and interestingly the backlash is *inversely* proportional to the gear ratio.

SM-Cyclo™ Drives

A similarly esoteric gear reduction assembly is the Sumitomo SM-Cyclo™ gear system. These reducers are intended for relatively heavy-duty applications, and offer very high efficiency at high-reduction ratios. For example, an SM-Cyclo drive with a ratio of 1:87 might offer an efficiency of 95% compared to an efficiency of 85% for a conventional three-stage reducer, and 75% for a worm gear. Additionally, the sinusoidal rolling action of this type of gear assembly almost eliminates wear and makes it very reliable.

There are at least two disadvantages to these units. They are somewhat heavy, and they become somewhat less efficient when run at power levels significantly below their ratings. This later problem is due to the stiffness or starting torque that is inherent in the design. Because of the importance of efficiency, the SM-Cyclo reducer is a good choice for the main drive reducer in a medium- to heavy-duty mobile robot.

CONCLUSIONS

The selection of the proper type of actuator and linkage for any given application is a complicated process. The choice depends heavily on the function that is expected of the robot in general and of the specific joint of the robot. New motor technology has greatly improved the performance of electrical servo systems, but still many applications require hydraulic actuators. Air motors are also finding significant use in fixed robots.

Although analog driver techniques are still very popular, the growing tendency is toward digital control. In the future, motor control amplifiers will be more and more intelligent, and as a result, the capabilities of machines will be more fully exploited.

REFERENCES

1. "DC Motors, Speed Controls, Servo Systems." Electro-Craft Corp., 1600 Second Street South, Hopkings, MN 55343.
2. "Permanent Magnet DC Servomotor Applications Seminar Handbook." Industrial Drives Division, Kollmorgen Corp., 201 Rock Road, Radford, VA 24141.
3. "Design Engineers Guide to DC Stepping Motors." Superior Electric, 383 Middle Street, Bristol, CT 06010.

MANUFACTURER REFERENCES

M1. Electro-Craft Corp.
     1600 Second Street South
     Hopkins, MN 55343
     (612) 931-2700
M2. Inland Motors
     Industrial Drives Division, Kollmorgen Corp.
     201 Rock Road
     Radford, VA 24141
     (703) 639-2495
M3. PMI Motors
     Division of Kollmorgen Corp.
     5 Aerial Way
     Syosset, NY 11791
     (516) 938-8000
M4. Superior Electric Co.
     383 Middle Street
     Bristol, CT 06010
     (203) 582-9561
M5. Motorola Semiconductors
     Box 20912
     Phoenix, AZ 85036
M6. Cybernetic Micro Systems
     445-203 San Antonio Road
     Los Altos, CA 94022
     (415) 949-0666

M7. Compumotor
1310 Ross Street
Petaluma, CA 94952
(800) 358-9068

M8. International Scientific Industries
Scattergood Drive
Christiansburg, VA 24073
(703) 382-1473

M9. Cincinnati Milacron
Industrial Robot Division
Lebanon, OH 45036
(513) 932-4400

M10. Bird-Johnson Company
110 Norfolk Street
Walpole, MA 02081
(617) 668-9610

M11. USM Emhart
Harmonic Drive Division
51 Armory Street
Wakefield, MA 01880
(617) 245-7802

M12. Sumitomo Machinery Corp. of American
Seven Malcom Street
Teterboro, NJ 07608
(201) 288-3355

M13. IMC Magnetics Corp.
100 Jericho Quadrangle
Suite 221
Jericho, NY 11753
or
IMC Magnetics Corp.
Western Division
12627 Hiddencreek Way
Cerritos, CA 90701
(213) 926-0927

M14. International Robomation/Intelligence
2281 Las Palmas Drive
Carlsbad, CA 92008
(714) 438-4424

Chapter 2

Manipulators

When the actions of several servomotors (as discussed in Chapter 1) are orchestrated through machine intelligence for the purpose of manipulating objects, the resulting mechanism is called a *manipulator*. Most of the early practical "robots" have been nothing more than semi-intelligent manipulators. The Robotics Institute of America, after long consideration, decided on the following definition of the term *robot*, "A robot is a programmable, multifunction manipulator designed to move material, parts, tools, or specialized devices, through variable programmed motions for the performance of a variety of tasks."

As the field of robotics matures, this definition of the word *robot* likely will be looked back on as being rather liberal. Because of this, there have already been attempts at defining more advanced mechanical entities with such terms as *android*, *cyborg*, and *mandroid*. Although there is little purpose in becoming involved in the genealogy of "mechanical beasties," it is useful to consider pick-and-place robots as manipulators or mechanical arms. Thus, a robot may consist of nothing more than a manipulator and a processor, or it may include several manipulators, a variety of elaborate sensor systems, a hierarchy of computers, and even a mobile platform.

TYPES OF MANIPULATORS

An enormous variety of practical robotic manipulators is available for a wide range of intended applications. To begin with, we will consider only *arm-type manipulators*. There are several ways to categorize these robots, including lifting capacity, application, and degrees of freedom. When arm-type robots are categorized by lifting capacity, the following general descriptions are used: *light duty*—15 lbs or less; *medium duty*—15 to 50 lbs; *heavy duty*—50 lbs and up. The capacity of the robot is normally the heaviest weight it can safely and routinely lift while its reach is fully extended.

Earlier, the term pick-and-place was used to describe all arm-type manipulators. Although this terminology is common usage in the industry, it also refers to a more specific subset of these robots. When robot manipulators are described by application, the term pick-and-place refers to those robots that are fitted with some sort of grasping mechanism (usually called a *gripper*) at the end of the arm. Most general-purpose robots are designed to accept a variety of such devices (called *end effectors*, *effectors*, or *tools*).

Thus, the application classification of a robot can often be changed by the simple replacement of its end effector with one of another type. Unfortunately, this "occupational" change of the robot may not be practical unless the robot was designed with the new application in mind. For example, a spray-painting robot *must* be explosion proof. The sparks from a brush-type servomotor could easily cause an explosion if no precautions were taken in such a robot. On the other hand, an arc-welding robot must be especially designed to keep the enormous induced electrical signals from interfering with its servo feedback and other inputs.

The software capabilities of the robot must likewise have been designed to support the application. For example, Cincinnati Milacron provides special "weaving" motions that may be specified in the program of the robot, to allow for very wide welds. [M1] Even more basic functions must be supplied in the programming of the robot. For example, in welding applications there must be a means of controlling the welding current and wire feed, in painting applications there must be a simple means of controlling the air flow through the spray nozzle, and in pick-and-place applications there must be a program function for opening and closing the grasping effector.

Finally, applications often require special sensors for optimum performance. For example, a wide variety of devices has been developed for the single function of following a seam during welding. Thus, it should become clear that although many manipulator arm-type robots may look the same, their interchangeability must not be taken for granted. The goal of most robot manufacturers is to address as many applications as possible with each model of robot that they make. Unfortunately, practical considerations often induce manufacturers to market their robots for fields of application in which the robots are not fully designed, equipped, or programmed. It is thus essential for the buyer of a robot to assure that the supplier has expertise not only in robotics, but also in the intended field of application.

As mentioned earlier, the third general method of categorizing manipulator arm-type robots is by their number of *degrees of freedom.* Each mode of motion of the end effector of the robot constitutes a degree of freedom. Actions of the end effector itself (such as grasping) are not considered as degrees of freedom. If the end effector is mounted on a "wrist," the modes of motion of the wrist are included in the total degrees of freedom of the robot. The six basic degrees of freedom are translation in each of the three axis directions and rotation normal to each of the three planes. Each joint of a robot may or may not add a degree of freedom.

The two robots shown in Fig. 2-1 are among the most common arm-types. The motion at each joint may be driven by hydraulic, pneumatic, or electrical actuators, as discussed in Chapter 1. Hydraulic systems usually use direct drive through cylinders or rotary actuators, and pneumatic systems often use air motors coupled through gear or chain reduction systems. Electrical servos usually use gear reduction systems although some research is being done on direct drive brushless dc motors.

DEFINING THE VARIOUS JOINT MOTIONS OF A ROBOT

The robot shown in Fig. 2-1A is a *straight-arm* robot with the following six joint motions:

1. Base (or arm) sweep
2. Shoulder rotation (also called arm pitch)
3. Arm extension
4. Wrist Pitch
5. Wrist roll
6. Wrist yaw

The linear "slide" action of the arm constitutes what is called a *prismatic joint.* Rotational degrees of freedom are referred to as *revolute joints.* In some applications, there may be no wrist at all, and a specialized end effector may be connected directly to the arm.

The robot of Fig. 2-1B is a *jointed-arm* robot with six joint movements:

1. Base sweep
2. Shoulder swivel
3. Elbow extension
4. Wrist pitch
5. Wrist roll
6. Wrist yaw

Like most robots, these two basic types are often made in a modular fashion. When wrists are present, they often do not allow yaw, or even roll. Several robots representative of these two fundamental types with various degrees of freedom are shown in Figs. 2-2 through 2-5. The Mini Mover robot of Fig. 2-2 is intended primarily for low-cost educational and experimental use.[M2] The other robots shown vary from the Westinghouse Electric Corp.[M3] light industrial assembly robot (Fig. 2-3) through the medium-duty Maker[M4] robot (Fig. 2-4), to the heavy duty Thermwood Corp.[M5] robot (Fig. 2-5).

POSITION CONTROL

At each joint of the robot, there must be some means of measuring position. This may be done with absolute or relative encoding (see Chapter 1). Relative encoding may be accomplished by the use of incremental motion servos such as stepping or brushless dc servomotors. The Mini Mover robot of Fig. 2-2 uses stepping motors in this way. Unfortunately, systems using open-loop incremental servos (such as stepping motors) can become unsynchronized if phase slippage occurs. Because of the dangers associated with loss of synchronization, this technique is only used on very light duty robots.

Relative encoding may also be accomplished by the use of a bidirectional pulse tachometer (see Chapter 1). This technique can offer very high resolution, and does not suffer from the slippage problems of open-loop incremental motion. However, all robots that use relative encoding must go through a *calibration* cycle when power is first applied. This cycle may occur automatically, or it may be caused by a statement in the program of the robot. During calibration, the robot will usually drive all servos slowly toward one extreme. Each joint (revolute or prismatic) has a limit sensor at the calibration extreme, and the motion of a servo will be halted when this position is

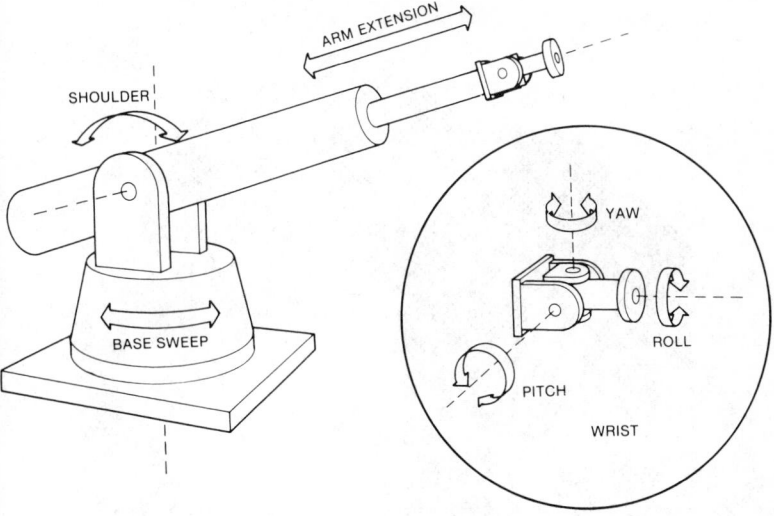

(A) A straight-arm robot.

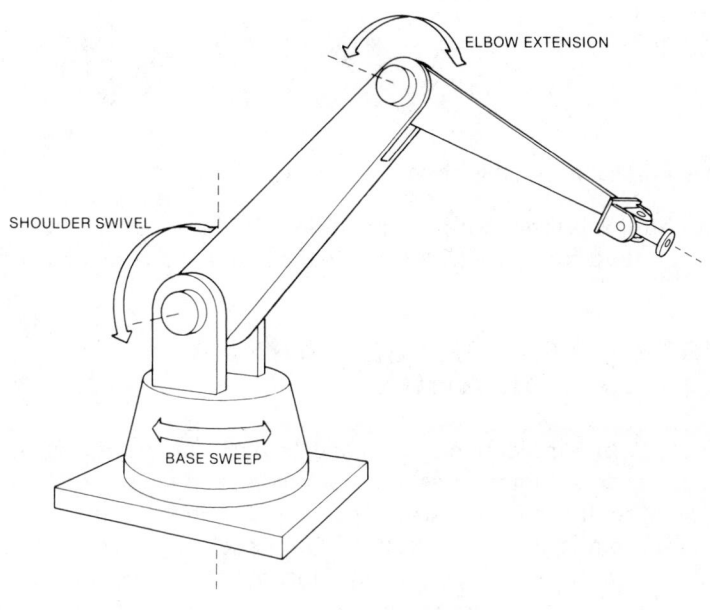

(B) A revolute-arm robot.

Fig. 2-1. Joints of a straight-arm and revolute-arm robot.

65

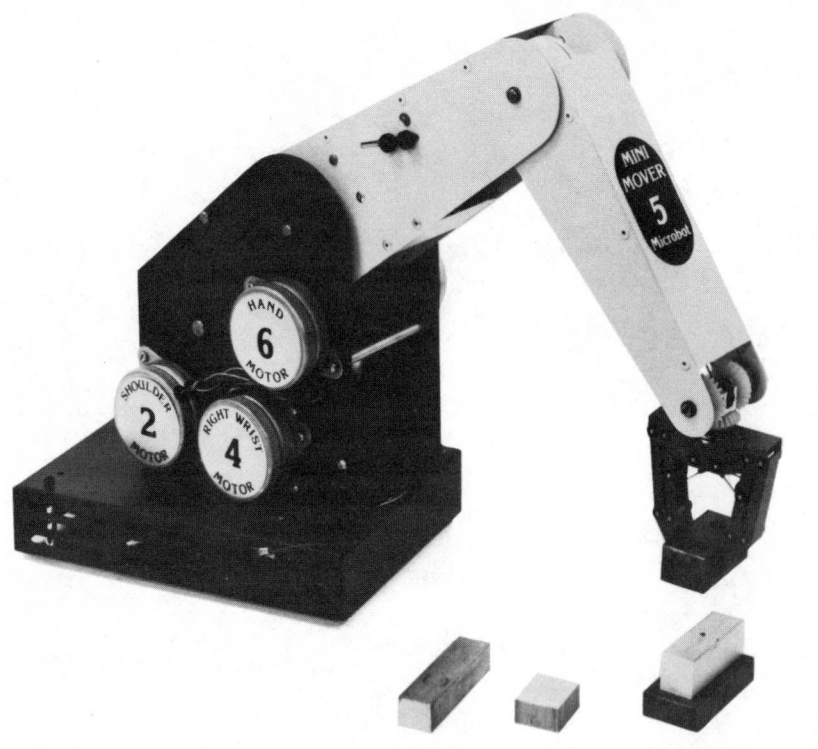

Fig. 2-2. A light-duty revolute arm.

reached. After this procedure, the position is continuously calculated from the relative motion with respect to the known starting position.

MOTION TRANSFORMATION AND COORDINATION

Before a manipulator can be used to perform real tasks, the motion of its various joints must be coordinated to each other, and to the frame of reference of the user. This coordination is usually done by one or more microprocessors. With the cost of *single-chip micro-computers* having come down to only a few dollars, many robots are being designed with a microprocessor for each joint. (Single-chip microcomputers are devices containing a processor, program memory, read/write memory, and interface ports on a single silicon wafer.)

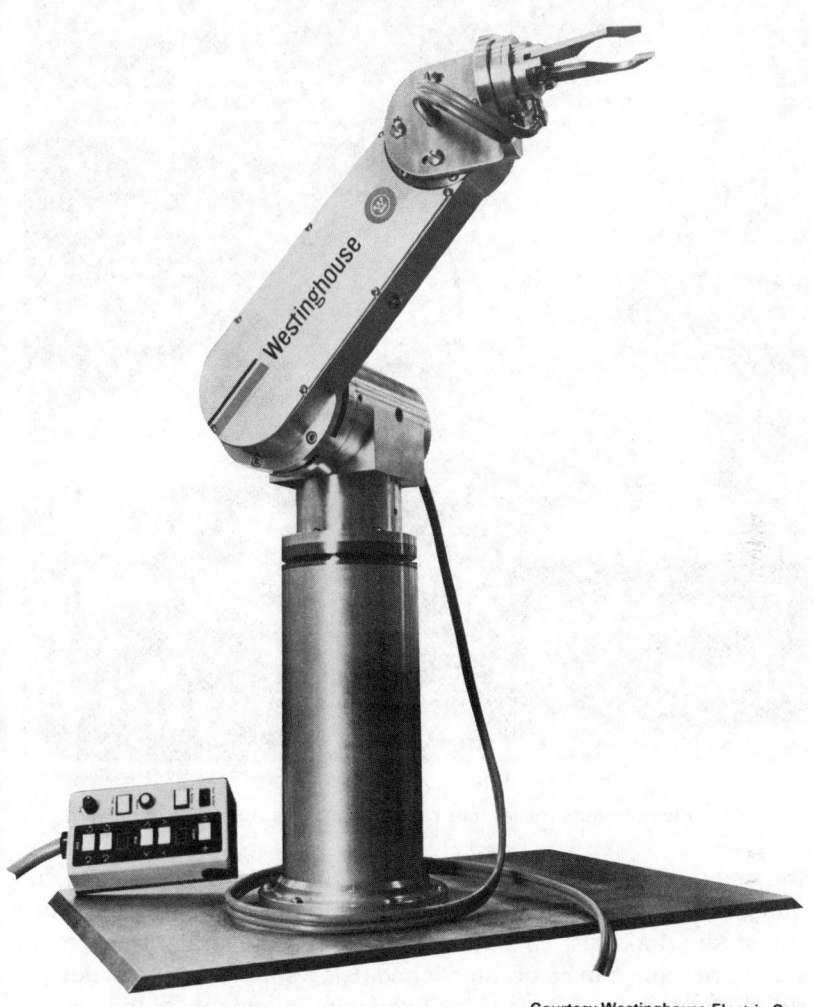

Fig. 2-3. A light-assembly robot.

Conversion of the mathematical description of a point, path, or motion, from one set of measurements to another is called a transformation. We humans have a great affinity for the Cartesian (rectilinear) coordinate system. Asked to describe the shape of an object, we will invariably give its height, width, and length. The robots of Fig. 2-6 do not share our perspective. For the straight-arm robot of Fig. 2-6A to place the end of its arm at a point, it must know the angles at which to position its shoulder and base-sweep servos (θ and ϕ), and

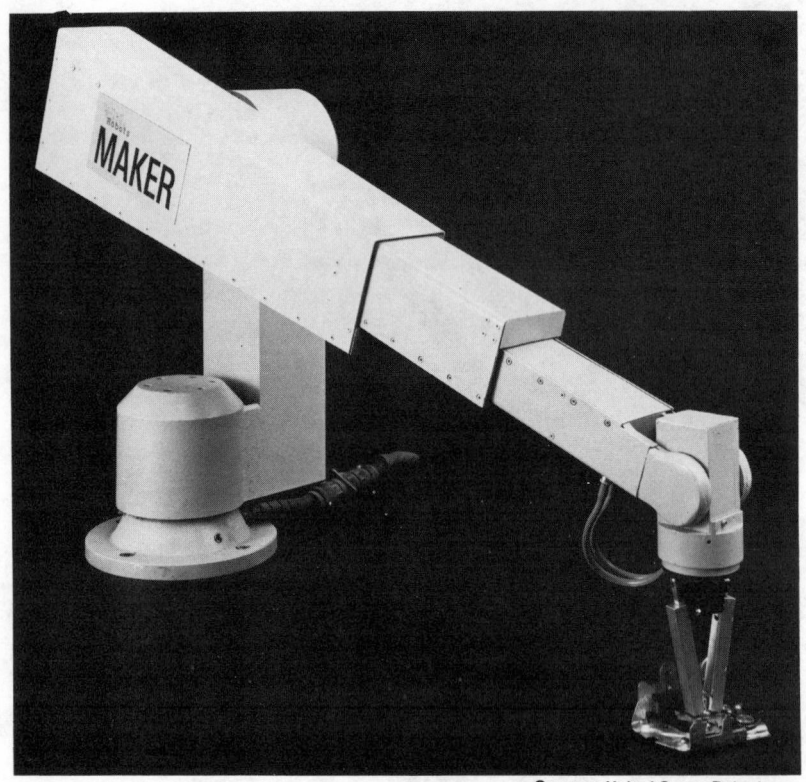

Fig. 2-4. A medium-duty maker robot assembling an automotive door latch.

the distance to extend its arm (d). In most robot programming languages instructions are entered in rectilinear coordinates, and a transformation must be done. In the case of the straight-arm robot, transformation can be accomplished using simple spherical geometry. Notice that we have not included a wrist and tool at this point.

The jointed-arm robot of Fig. 2-6B operates in what are called *revolute coordinates* because the reach is determined by a revolute joint instead of a prismatic joint. This coordinate system requires a somewhat harder transformation. One approach is for the computer to calculate the polar coordinates as if it were driving a straight-arm robot. The distance (d) to the desired point is thus the base of a triangle composed of the outer- and inner-arm elements. Since the lengths of these elements are constant and known, the three sides of a triangle can be found, and the interior angles can thus be found. In practical, real-time, computer programming the classical solution of

Fig. 2-5. A heavy-duty five-axis robot (50-lb load, 0.060-inch repeatability).

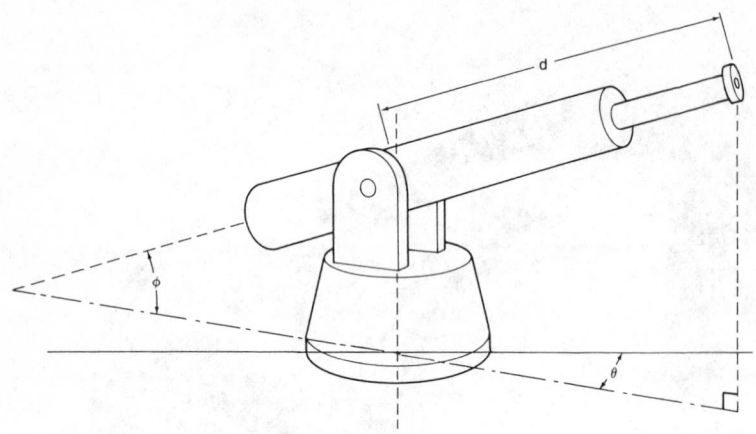

(A) A straight-arm robot.

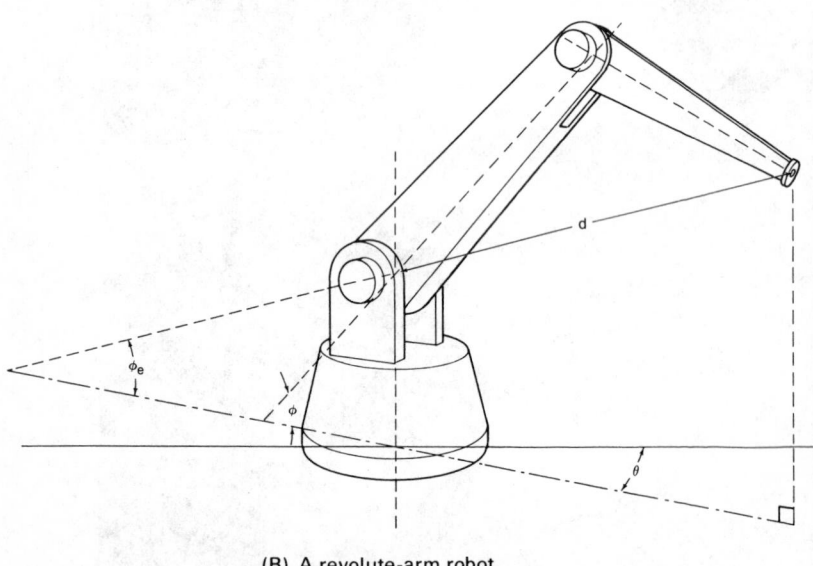

(B) A revolute-arm robot.

Fig. 2-6. The spherical coordinate system of the straight-arm robot and its equivalent for the revolute arm.

these types of problems may require too much computer time. A common alternative is to provide a table of elbow extension angles, versus the distance (d). To increase the resolution of such solutions, a simple interpolation routine is often used. With these techniques, the computer can rapidly determine the elbow extension angle and, thus, the actual shoulder swivel angle.

Up to this point, the solutions have been rather simple. We have been assuming that we were given the position of the base of the wrist. Unfortunately, this is usually not the case. In most instances, we are interested in the point of contact of the tool. We may also have constraints on the angle of the tool with respect to the work. Because of this, usually the angle of the tool with reference to the "world," can be found, and then the problem can be worked backwards to find the point at which the base of the wrist must be positioned. The positions of the arm servos are then calculated (as already discussed). Once the angles of the arm (or the outer element of the arm) are known, they may be subtracted from the "world" angles of the tool to find the wrist servo angles.

COORDINATED MOVEMENT

Given that a robot is to move between two positions, it might first calculate the servo settings for the new position. The relative *average* velocity of each servo would thus be proportional to the difference between the current position of the servo and the position to which it must move. If the absolute speed of the tip of the tool is to be accurately controlled, the relative speed of each servo must be scaled by the straight-line distance of the movement (at the tip of the tool). This can be stated mathematically as

$$Vn = Va(Pn - Pn')/ Da$$

where,

Vn is the average velocity of servo n, measured in linear motion at the end of the arm,
Va is the average absolute linear velocity of the end of the arm,
Pn is the destination position of servo n,
Pn' is the present position of servo n,
Da is the absolute linear distance to be traveled by the end of the arm.

Since the absolute linear distance to be traversed by the end of the arm (Da) may lie along any vector in space, it must be calculated by taking the square root of the sum of the squares of the distances along the three principal axes.

Unfortunately, the servo velocity that we calculated above was the *average* velocity, and the command to the servo amplifier is a steady-state target velocity. Fig. 2-7 illustrates this problem. Notice that two servos are moving at constant velocity through point 1. When point 2 is reached, the path deflects in such a way that servo 2 must come to a new velocity, while servo 1 remains at constant speed. Unfortunately, this new velocity cannot be reached instantly because the servo must accelerate the arm in its plane of control. During the time that servo 1 is maintaining its velocity, and servo 2 is accelerating (the shaded area in Fig. 2-7), servo 2 will lose ground relative to servo 1. In this example, servo 1 would be in proper position at point 3 before servo 2, and the motion of the arm would be off of the intended path.

The computer could correct this acceleration problem in one of

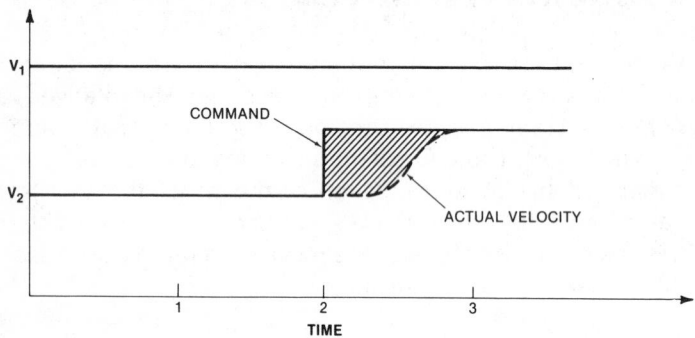

Fig. 2-7. Servo response for uncompensated commands.

several ways. For example, it could divide the path between points 2 and 3 into two segments (Fig. 2-8). During the first segment, the velocity command would be larger than the desired average velocity by an amount that would exactly bring the average velocity for both segments to the desired level. By doing this, the path between points 2 and 3 will be slightly warped, but both servos will reach point 3 at the correct time and with the correct velocities.

Now, to reduce the warp in the path between points 2 and 3, the computer can take the original path and divide it into several smaller paths. These subpaths could then be divided in two and treated as discussed earlier. Unfortunately, the acceleration of the servos will be determined by the load carried by the arm and the position of the arm. Thus, it is important to make calculations at very small intervals so that these linear approximations will be relatively accurate. What this means is that we are moving the problem to smaller and smaller subpaths. This is essentially a *differential* treatment of the problem. (Keeping this in mind will help clarify the concept behind Jacobian transformations when we discuss them.)

By now, the reader is probably beginning to appreciate the potential complexity of the control problem. However, several trade-offs can be made to avoid these complexities. For example, avoid the need for coordinate transformation by simply having the programmer lead the arm between path points with a hand-held control box. Such controls are referred to as *pendants*, *umbilical*, or *ICMs* (*Interactive Control Modules*). The operator simply moves the arm servos until the tool reaches the desired point, and then enters the point into the *path program* of the computer, along with velocity and other information.

When a robot programmed in this way is commanded to execute the program, it simply plays back the spatial recording. The computer of the robot may subdivide and manipulate the path commands (as described earlier) to reduce path distortion. The accuracy with which the robot will track the desired path is a function of the number of points saved, the control algorithm, and the power and accuracy of its servo systems.

In many applications (such as seam welding), the robot must follow a complex curve with extreme accuracy. If one is to optimize such systems, a concise, comprehensive mathematical representation is needed.

MATRICES AND JACOBIAN TRANSFORMS

The reader who is familiar with three-dimensional computer graphics will already appreciate the power offered by homogeneous matrices. By relatively simple manipulation of elements in matrices, it is possible to translate points, lines, planes, and even forces from one coordinate reference frame to another. This translation is precisely what is required to add order to the computational quagmire just described.

The wrist of the revolute arm shown in Fig. 2-9 is mounted to the end of the outer-arm element. Therefore, all of the servo commands to this joint must be with respect to this *reference frame* (which we will call frame 3). The position of frame 3 with respect to frame 2 can be described by a matrix transform that we will call A3. Frame 2 is likewise referenced by the transform A2 to coordinate frame 1.

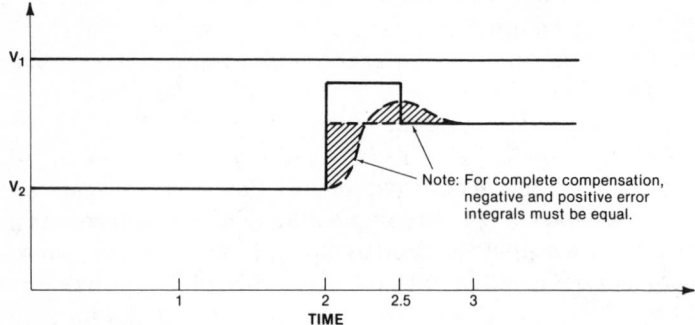

Fig. 2-8. Servo response for compensated commands.

Finally, Frame 1 is referenced to the "world" (frame 0) by transform A1.

Jacobian matrices and inverse Jacobian matrices provide a direct transformation between a coordinate frame and the world in terms of *differential* changes. The Jacobian that referenced frame 3 to the world (frame 0) would be referred to as the T3 Jacobian. Given the present positions of the joints of the robot, and the desired differential changes in the joint angles, the T3 Jacobian would provide the differential changes in position of the end of the arm in the real-world frame of reference. Conversely, the inverse T3

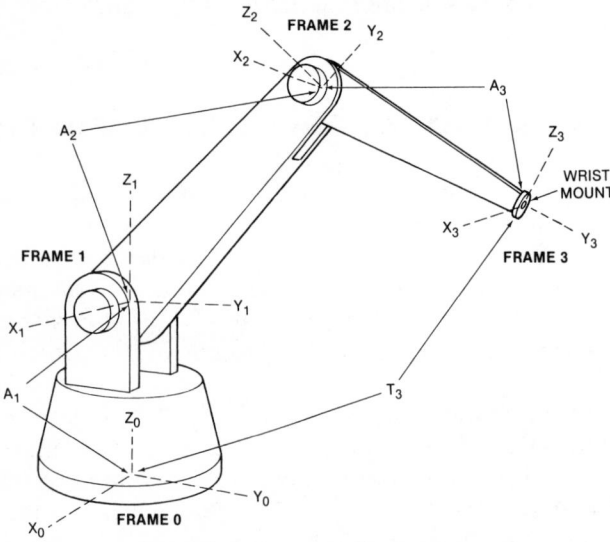

Fig. 2-9. Transforming coordinate frames.

74

Jacobian transform would define the differential joint-angle changes that would be required to move the end of the arm through a specified X, Y, and Z differential movement (in the real-world frame of reference), given the present X, Y, and Z position.

The subject of matrix mathematics and Jacobian transforms is far too large to cover in any serious way in this book. Paul, in his excellent book, *Robot Manipulators*, presents a concise treatment of the subject, and the seriously interested reader is urged to read this work. Unfortunately, because of the computational speed of most control computers, it is not usually possible to fully implement such theory in real-time applications.[2] This implementation is not possible because of a conflict between the rate of servo sampling required and the computational speed of the processor. One compromise is to calculate the Jacobian at every tenth sample and to use simple interpolations in between. With the advent of more and more powerful microprocessors and high-speed math chips, the full implementation of Jacobians will appear gradually.

FORCE SENSING AND COMPLIANCE

Thus far, we have considered the motions of the robot as a sequence of absolute position commands. To understand the weakness of this simplification, consider a robot that needs to pick a small ball bearing off of a table surface (Fig. 2-10). Unfortunately, the

Fig. 2-10. The application of compliance.

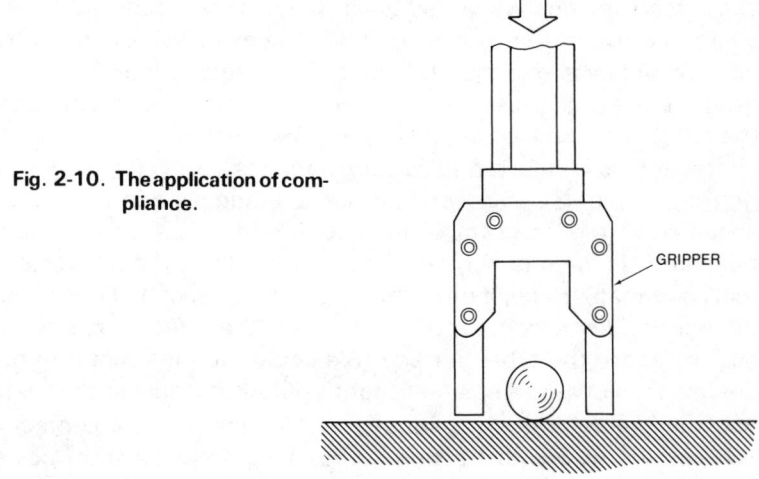

GRIPPER

surface is not perfectly uniform in height. When the arm lowers the manipulator around the bearing, it would be desirable if it just lightly touched the table surface. If it does not touch the surface, it may not properly grip the bearing. If it pushes too hard against the surface, it may jam the action of the effector (gripper). The effector could be equipped with a touch switch, but this would not provide a measurement of the amount of force being exerted. Alternatively, the effector could be equipped with a pressure sensor, or the force being exerted by the downward driving servo could be measured, or both.

The force being exerted by a servo can be determined in a number of ways. For an electric servo it can be gauged roughly by the current that the motor is drawing. Unfortunately, the measurement is influenced by frictional losses, temperature effects, the position of the arm, and other factors. (Even if the current is not used for this purpose, it should be monitored for the detection of fault conditions.) Similarly, the force being exerted by a hydraulic actuator can be determined roughly by the pressure driving the actuator. Alternatively, strain gauges mounted at special positions along the arm members can be used to more accurately determine force.

When force is sensed at any place other than the end of the effector (or wherever it is to be controlled), it must be remembered that the forces on the effector are only indirectly related to the servo forces. Thus, we have another application for transforms.[2]

Returning to the example of the robot and the bearing, we would have an additional programming problem. To press the effector against the table at a force of 3 units (using only position commands and force sensing), we would have to generate a control sequence something like that shown in Fig. 2-11. The robot would approach to a position safely above the table and incrementally move downward until the required pressure was reached. It would be much better if the robot had a built-in ability to trade off position and force. Such a characteristic is referred to as *compliance* (see also legged locomotion in Chapter 3). In advanced robots, a compliance force can be specified in the program of the robot, and it can be changed as needed. This force is normally specified for the various axes of the tool coordinate frame. If our bearing-grabbing robot had this capability, we could simply command it to place the effector at a position slightly above the table surface. We could then instruct it to move downward slowly to a position slightly below the table surface with a specified vertical (Z) compliance force. At the point of contact (restraint), the downward acting servos would switch from *position servos* to *torque servos*. This has an added advantage. If someone

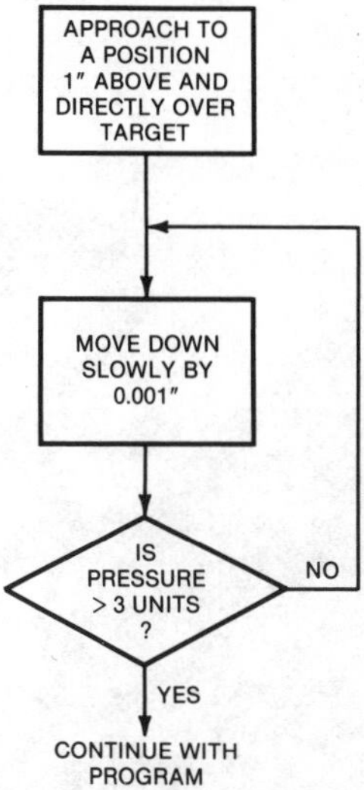

Fig. 2-11. The problem of approaching to contact without compliance.

```
┌─────────────────┐
│  APPROACH TO    │
│  A POSITION     │
│  1″ ABOVE AND   │
│  DIRECTLY OVER  │
│  TARGET         │
└─────────────────┘
         │
         ▼
┌─────────────────┐
│  MOVE DOWN      │
│  SLOWLY BY      │
│  0.001″         │
└─────────────────┘
         │
         ▼
      ╱ IS ╲
     ╱PRESSURE╲    NO
    ╱ > 3 UNITS ╲ ─────→
    ╲    ?    ╱
     ╲      ╱
         │ YES
         ▼
   CONTINUE WITH
   PROGRAM
```

removed the table, the robot would not attempt to move all the way to the floor, but would stop at the position commanded.

Another example of compliance is the problem of inserting a peg into a hole. If more than a slight pressure is exerted against the walls of the hole, the peg might jam. For a vertical hole, this would ideally require a zero-force compliance in both the X and Y directions. In reality, it is not possible to comply at zero force, but quite small forces can be achieved.

Compliance is desirable both for ease of control and safety of operation. For safety, most compliance algorithms have built-in force limits. It should also be noted that rotational (torque) compliance is also possible although seldom implemented. Torque compliance is very useful in screwing and unscrewing operations.

GRIPPERS AND TOUCH

One of the most common types of end effectors is the gripper. Since the pick-and-place market is variously estimated to be between 25% and 35% of the total (arm-type) robot market, considerable attention has been given to gripper requirements. A variety of grippers is shown in Figs. 2-12 through 2-14. Most grippers offer passive compliance in the gripping action. Although active compliance (as already discussed) requires computations on the part of the control computer, passive compliance does not.

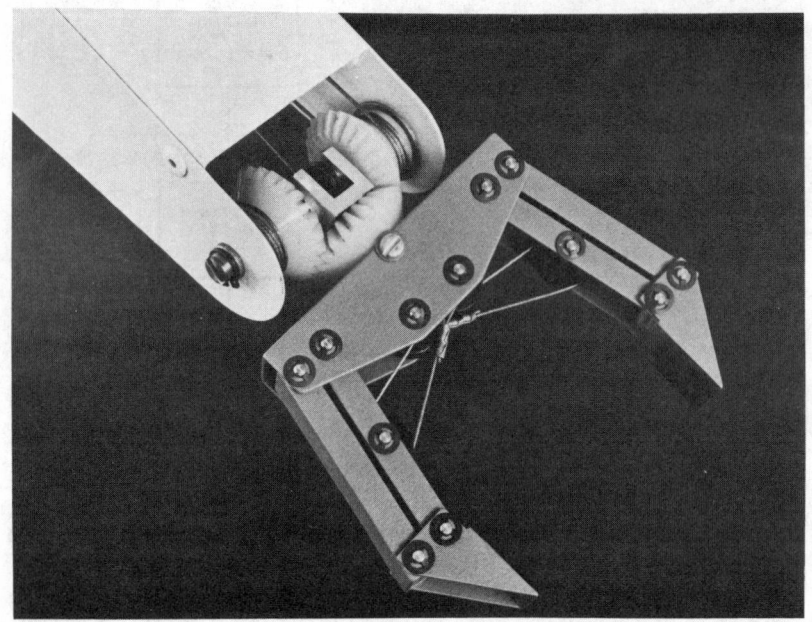

(A) Photograph of gripper.

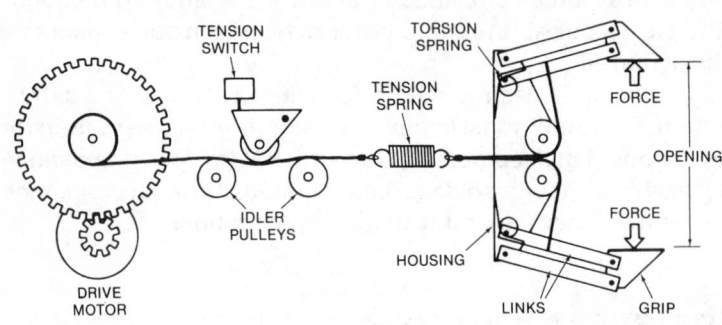

(B) Diagram of gripper.

Courtesy Microbot Inc.

Fig. 2-12. The gripper of the Microbot Mini Mover 5.

The gripping action for the Mini Mover unit (Fig. 2-12A) is driven by a cable as shown in Fig. 2-12B. Notice that passive compliance is provided by a tension spring. This is especially important because the gripper is powered by a small stepping motor. Mechanical compliance thus reduces the danger of phase slippage in the motor as well as reducing the control problem for the computer. A tension switch is provided to signal the computer that the gripper is firmly closed. Notice that the jaws of the gripper are designed to close in a parallel manner, thus avoiding any outward force that might cause the targeted object to be expelled.

The industrial grippers of Fig. 2-13 are designed for a variety of different-sized target objects. These grippers, like most industrial grippers, are pneumatically driven. Because air is compressible, it provides a certain amount of passive compliance. Simple and inexpensive pressure transducers can be inserted in the feed lines of such grippers to monitor the gripping force. Although this may be sufficient for many applications, it is often necessary to have separate force sensors on each "finger" of the gripper. In this way, the hand position can be corrected if one finger contacts the target before the other.

Courtesy Mack Corp.

Fig. 2-13. B•A•S•E™ pneumatic grippers.

Fig. 2-14 shows another interesting feature. Here the gripper is mounted on an *X-axis transport*, which is simply a prismatic joint that is pneumatically driven. Using such a device, the robot can make its final approach on the target without operating its main servos. This device makes the job of the computer much easier by providing a

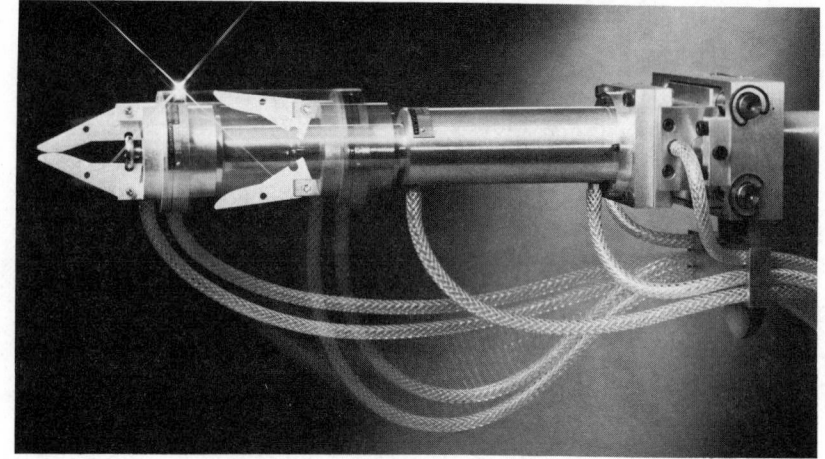

Fig. 2-14. B•A•S•E™ X-axis transport with pneumatic gripper.

degree of passive compliance without the need for complex transform calculations.

In addition to the normal mechanical considerations, grippers must often be designed to perform under very adverse conditions. Moving parts into and out of caustic solutions, furnaces, and other unpleasant places (Fig. 2-15) are jobs that workers and management readily concede to robots.

Although the sensing of collective or differential gripping force (as described earlier) is often adequate, the grasping of oddly shaped objects may require more dexterity. This can be provided by using pressure-sensitive materials over the gripping surfaces. By dividing these surfaces into a grid, the computer can sense the balance of force over the contacting surfaces.

In one experiment objects such as nuts, bolts, flat washers, lock washers, and other items have even been identified by touch.[3] In this experiment, Hillis used a sensor array composed of a flexible printed-circuit board and a sheet of anisotropically conductive silicone (ACS) rubber. ACS rubber has the convenient property of being electrically conductive only along one axis in the plane of the sheet. The ACS rubber sheet and the flexible printed board were separated by a nylon mesh (known to the layman as a nylon stocking) as shown in Fig. 2-16. As increasing pressure was applied to an area of the ACS sheet, more of the sheet would come in contact with the flexible printed-circuit board, and the resistance in that region would go down. The flexible pc board was etched with parallel conducting lines

Fig. 2-15. A Unimate robot processing hot castings.

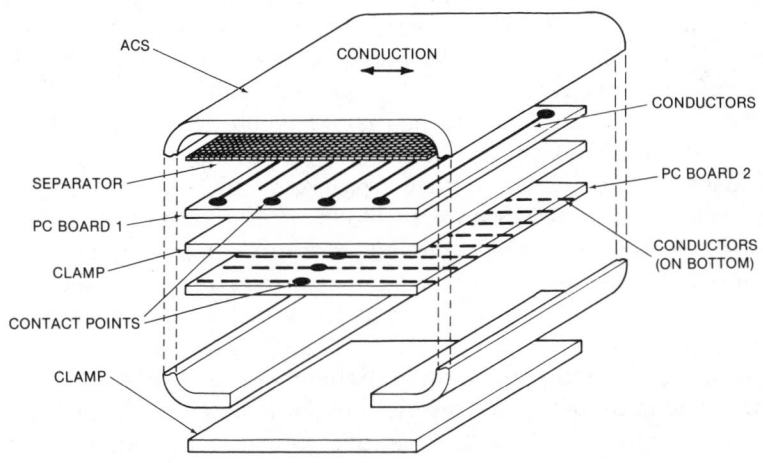

Fig. 2-16. The Hillis-touch sensor.

at 90° to the conductive direction of the ACS sheet. A second circuit board was used as a contact board for driving columns along the conductive lines of the ACS sheet. The result of this configuration was a sensing grid with a resolution of approximately 0.01 square centimeter (approximately that of the human fingertip).

Data from such sensors can be treated very much as is image data from a camera (see Chapter 4). As pointed out by Hillis, this data is in some ways easier to process. For one thing, there is less tendency for background interference because the image is confined to the plane of contact. This is somewhat like having a camera with an *extremely* narrow depth of field. The robot may also better define the image by applying more or less pressure.

Unfortunately, there are some problems with touch. One problem is that the sensor may be more easily damaged than unsensitized gripping surfaces. Furthermore, pressure-sensitive materials tend to have hysteresis and memory effects that could become troublesome in constant daily use. Besides ACS, a number of other materials with resistive and piezoelectric properties are being explored for this application. Simple, fairly rugged touch sensors likely will be an integral part of many future grippers.

PROGRAMMING THE ROBOT ARM

As mentioned earlier, many robots are programmed by walking them through the desired movements using a *teaching pendant*. *Pendant programming* usually allows some primitive editing. For example, the programmer may be allowed to back up through the program and modify path end points or speeds. When the path has been taught, it may be stored on magnetic tape (or disk). Additionally, the user may be allowed to load a program, single step through it, modify it, run it, and even store the modified program as a second file.

In the course of programming an arm with a pendant, some instructions (besides position and speed) may be accepted by the control computer. These instructions may include delays, such as wait X seconds before continuing or wait for a signal input before continuing, or conditional path branching. (More will be said about path branching later.) The more instructions included on the pendant, the more confusing it becomes to use, which means the teaching operation becomes more and more like a language, and the pendant does not offer the convenience of a keyboard and a CRT (cathode ray tube). What is needed for these more complex applications is a special computer language that incorporates both the standard features of a language and special motion control commands. There have been (and still are) many such languages.

Among the earlier languages was WAVE[2] and its successor AL. These languages calculated Jacobian transforms in advance, and unloaded the run time portion by storing the data in modifiable files. Unfortunately, they did not allow for easy path modification at run time (such as relative positioning that will be discussed later).

One of the best-known languages is called *VAL*. This was one of the first robotic languages to operate on a line-by-line basis at run time. Since VAL, many robotic languages have been introduced. The following list contains some of the languages in use at this time, along with the companies that offer them.

VAL	Unimation Inc.[M6]
RAIL	Automatix Inc.[M7]
Pascal	CGA/PLR[M8]
	Hodges Robotics International Corp.[M9]
LPR	Cybotech Corp.[M10]
ERG	Hodges Robotics International Corp.[M9]
EIA	Ikegai America Corp.[M11]
ISO	Ikegai America Corp.[M11]
BASIC	ISI Manufacturing Inc.[M12]
	Mack Corp.[M13]
	Microbot Inc.[M2]
	Mobot Corp.[M16]
	Robotic Sciences International Inc.[M14]
FORTRAN	Sigma Sales Inc.[M15]

Many of the newer robotic languages are derivatives of either BASIC or Pascal although they may not go under those names. The additional functions needed to drive a robot from an existing high-level language can be divided into two categories: an extended set of commands and an interrupt-driven servo control program.

The interrupt-driven program will execute at precise time intervals. When it executes, this program will pick up *set-point* commands for velocity and/or torque from memory locations where the main program left them. The interrupt program will also read the servo positions, calculate new drive level commands (see Chapter 1), and read other inputs. Information associated with these operations will then be left in memory locations where the main program will have access to it. A set of additional memory locations is usually used for *handshaking* between the main and interrupt programs.. The use of handshaking assures that one program will not use partially updated values from the other program. The interrupt program may also calculate new control algorithms or Jacobian transforms, or it may activate other tasks to do these (and other) functions. The interrupt program may thus be a simple single-timed interrupt, it may be a set of prioritized interrupts, or it may constitute a full multitasking executive.

The extended command set for the main program may be added in two ways: it may be assembled into the object code of the language, or it may be added to the applications program as a library of procedures (subroutines). When these functions are assembled into the language, programming is usually simplified, and additional safeguards can be installed. Most commercial robots that offer language control also offer a pendant to augment the language. Statements in the language often allow interaction between the language and the pendant.

The integration of the teaching pendant and the language is done in several ways. In the Cincinnati Milacron system (Fig. 2-17), points defined at the teaching pendant are transformed into rectilinear coordinates for storage.[M1] At this level, they are accessible to, and compatible with, the instructions entered at the terminal (keyboard, CRT). During execution, the coordinates are transformed back into the required revolute coordinate form.

Most advanced languages offer several operational modes, including editing, single-step execution, normal execution, and training. The better these functions are integrated, the more easily and safely the robot can be programmed. For example, if some operational problem (or the operator) should cause the normal execution of the robot to abort, it would be very helpful if the programmer were given the location in the program where the problem occurred. It would be even more helpful if a short *history* (sometimes called a trace stack) was available, giving all data concerning the last few actions of the robot prior to the halt. Especially in programs with a lot of branching and relative motion, this information can be extremely helpful.

Like all languages, the robotic control language should offer the following functions:

- Program and data storage
- Console and printer I/O (Input/Output)
- Arithmetic operations
- Logical (Boolean) operations

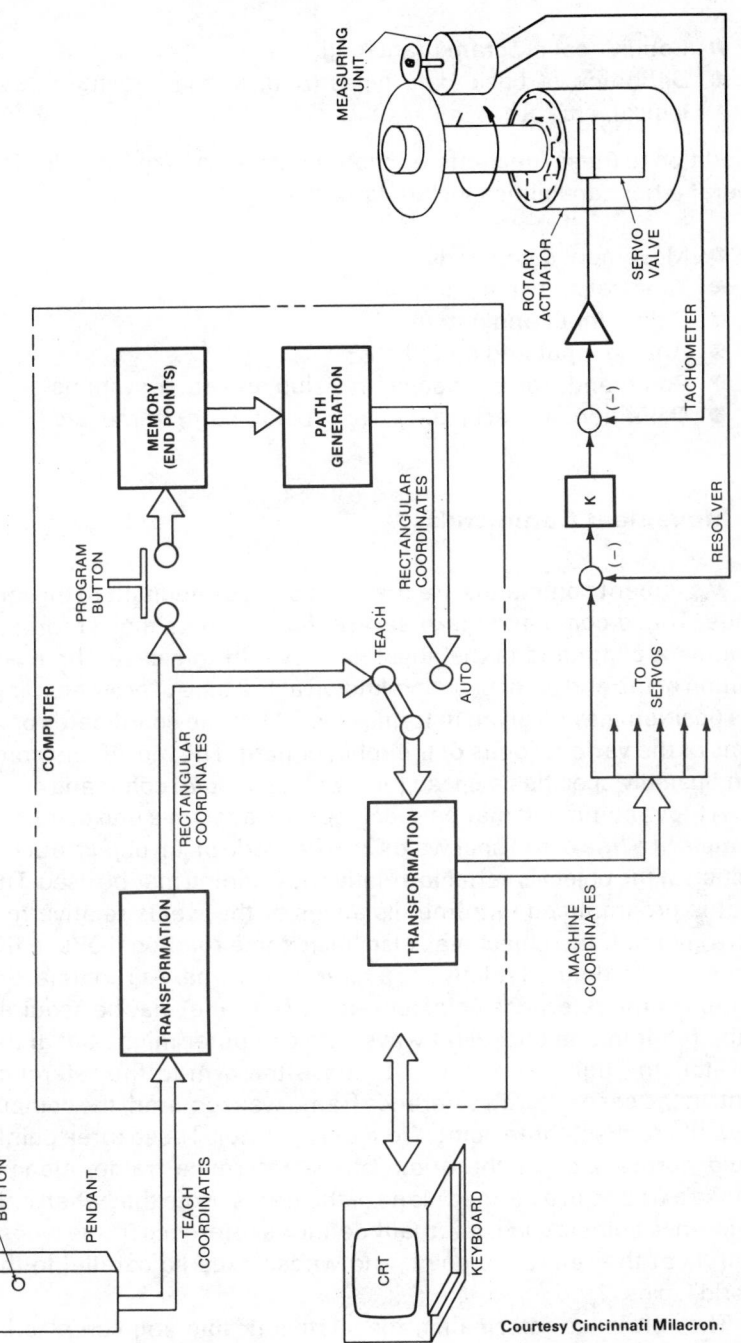

Courtesy Cincinnati Milacron.

Fig. 2-17. The integration of a control pendant with terminal and computer.

85

- Conditional program branching
- Definition of basic data types (real, integer, string, arrays, logical, etc.)

In addition to these functions, a robotic control language should offer several other capabilities including:

- Movement commands
- Time-related commands
- Digital input and output
- Analog input and output
- Commands for interfacing to vision and other systems
- Definition of special data types (point, path, frame, etc.)

Movement Commands

Movement commands are the most obvious addition to the language. These commands take several forms. The simplest form of movement command is the absolute move. In this case, the exact location of the end point of path is known at the time of programming. This position may be given in terms of world frame coordinates or in terms of the various joints of the robot or both. The movement command usually specifies a speed (or *speed-schedule*) command.

A move command may be specified with a *relative* end point. For example, if a line of ten spot welds is to be made on an object, but the position of the object is variable, relative positioning may be used. The robot is programmed with the distances of the welds relative to a known point. If the object may also be at some rotational offset, the points may be defined relative to a *reference frame.* At run time, the position of the reference point (or reference frame) may be acquired by the robot in one of several ways. The computer might signal the operator (through the pendant) to move the arm to the reference point. In the case where a reference frame was required, the computer would request three points from the operator. These three points would normally be at the origin of the reference frame, along a defined axis and in a defined plane of the frame. Note that when only a reference point is used, it actually defines a reference frame whose origin is at the reference point, and whose axes lie parallel to the "world" axes.

The reference information for relative positioning can also be acquired in other ways. The computer could be made to request position information from a vision system (see Chapter 4) or from a simple beam-type-object edge sensor. One common technique with conveyor systems is to equip the conveyor with a pulse tachometer (see Chapter 1) to measure its relative motion. The pulse tach is connected to a counter (implemented in hardware or software). At the instant an object moving down the conveyor breaks a light beam, the counter is reset. From that point until the next object breaks the beam, the counter will contain the position of the object relative to the light beam. Thus the count can be used directly as a relative position. Any points on the object (with respect to the leading edge of the object) would be specified as being relative to the light-beam position offset by the conveyor position.

In the case where a relative position reference is changing (such as the conveyor example), a relative position move does not end with the arm stopping. Instead, the arm will move at the same speed as the object until the next move is executed. In other words, the arm will remain motionless *relative* to the moving object as long as the object remains in reach.

Other examples of the usefulness of relative motion are stacking and unstacking operations. During stacking operations, a robot must move to a slightly higher position each time it returns to place an object on the stack. The vertical offset can be determined by adding the height of the objects being stacked to the last known stack height. The new position can then be used as the offset for a relative move. Of course, the converse is true for unstacking operations. If the height of a stack is unknown, or if the objects being stacked vary in thickness, compliance may be used to find the top. In an unstacking operation, for example, the robot might be programmed to go to a position directly above the stack and move downward slowly with a minimal compliance force. When the top of the stack was detected, the robot would read its position and use it for the relative moves that followed.

A robot may also be called upon to take one of several paths, in accordance with conditions encountered. There are two basic ways that this can happen: *fixed branching* and *on-condition branching.* In fixed branching, a statement is placed in the program that calls for a test. When this statement is reached, the computer will perform the test and take one of several program branches. These program branches will in turn cause the robot to take one of several physical paths. The test might concern the condition of an external input, or it might concern the value of some internal variable (for example an operation counter).

On-condition branching can take place anytime that a specified condition occurs. This condition has the same criterion as the test just described for fixed branching. The difference between on-condition branching and fixed branching is that the test is performed on a continuing basis by the computer, as long as the on-condition branch is left active. When a branching action is no longer appropriate, a statement in the program should deactivate it. An example of on-condition branching would be a tool cleaning maneuver that had to take place under a given condition. An on-condition branch may act as a subroutine returning control to the main program at the point where it was interrupted. It may also terminate the present operation. In this type of branching we do not know in advance where the arm will be at the time the branch occurs. For this reason, the location is often read as the first operation of the on-condition branch program. This position information is valuable, both for returning to the position, and for relative motion during departure. So common is the problem of approach and departure that some languages offer simple statements to facilitate the process. This will be discussed more when we look at the RAIL language.[M7]

Time-Related Commands

Most high-level languages were not written with real-time-control applications in mind. For this reason, they have either no time related commands or at most a simple WAIT command. It is essential that a robotics language have at least this simple WAIT instruction. This instruction allows the robot to wait a sufficient time for unsynchronized physical operations to stabilize. (Unsynchronized operations are those that take a significant time to accomplish but that do not signal their completion to the robot.)

Although the WAIT command is essential for robotic applications, more sophisticated timing functions are desirable. One such function is available in systems having multitasking capabilities and is called by such names as *TIMER* or *TICK*. When timers are implemented in a multitasking executive, there are usually several of them. The timers may be designed to count up or down. Down counters can be imagined as simple stop watches that alarm when they have reached a zero-count condition (timed out). A TIMER can be used to cause relative on-condition branching motion after a certain amount of time. It can also serve to limit operations when the robot becomes stuck. For example, if a robot was supposed to command a synchronized function (one whose completion is signaled to the robot), a

TIMER function could abort the process if it was unsuccessful after a predetermined period of time.

For example, if objects were coming down a conveyor line in groups, the robot could be programmed to watch for a gap of say 8 seconds or more between objects. When this interval is observed, it could cause the robot to branch into another function (such as packaging the group of objects just received). Unlike performing a WAIT instruction, the robot itself is free to continue executing its program while the timing function is being accomplished.

Digital Input and Output

As mentioned earlier, it is usually necessary for the robot to have the capability of controlling and monitoring devices other than its own servomotors. These signals can be divided into three groups: tool related, operator related, and external equipment related.

Tool-related signals include those that activate a gripper and read back its closure force, or those that are used to operate a welder. Since most tool-related signals are planned by the manufacturer of the robot, their control is often built into the language. For example, the language may support a gripper command such as OPEN. This command might do nothing more than setting a certain bit of an output port to a logical one, but it is much clearer than a statement like OUT PORTA, O1 H. When a nonstandard tool is to be used, it may be interfaced through spare I/O (input/output) port lines. When this interface is done, procedures (subroutines) may be defined for its operation. In languages such as Pascal, Ada, PL/M, and C, the procedure may be given a name (e.g., EJECT or DRILL _ ON), which makes its use very convenient. Most BASIC languages are line number oriented and thus do not offer this advantage.

Operator-oriented signals include programmable switches and panic buttons. The operator may, for example, be given several switches whose meaning is defined by the programmer. These switches can be programmed to cause fixed or on-condition branching. An example would be a switch that returns the arm to a home position for tool replacement. When the replacement was done, the work could continue from where it was interrupted.

External equipment signals come in a wide variety. Among the most common types of external signals are those for synchronizing the robot to a conveyor belt (as previously discussed) for controlling tool carriers and movable work tables (Fig. 2-18) and for handshaking

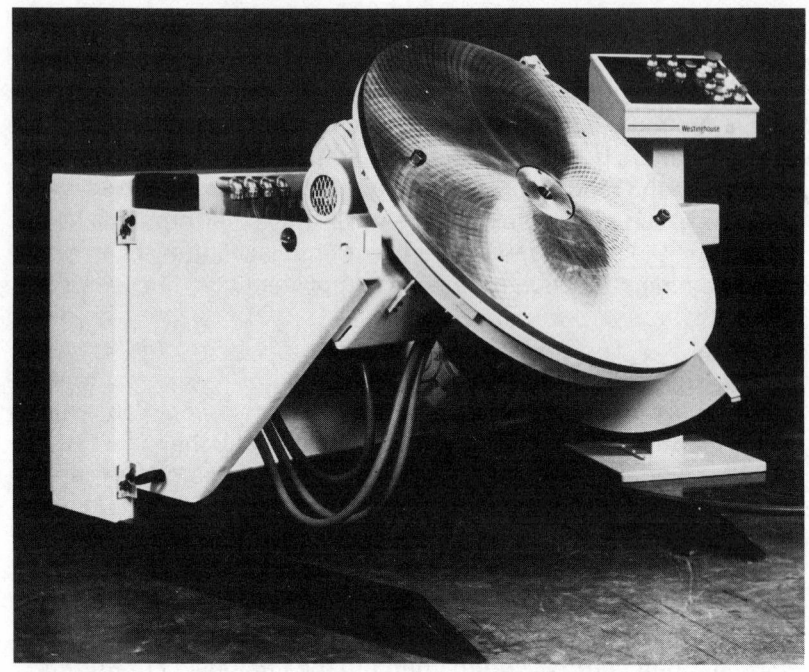

Fig. 2-18. A work-piece positioner (table).

with other robots. Most of these signals can be interfaced through a few general-purpose parallel I/O ports.

In addition to parallel I/O ports, serial ports are often available on robots. These ports are used to allow remote operation and monitoring of the robot and for diagnostic purposes. Troubleshooting of some robots can even be done over the telephone using a serial interface and a modem.

Analog Input and Output

Most industrial robots have the capability of reading (and sometimes putting out) analog signals although these interfaces may or may not be available to the user. Analog inputs are useful for reading temperature, pressure, and other transducer signals. Analog outputs are less common because many analog-type output functions can be more efficiently provided by using pulse-width modulated digital signals.

Commands for Interfacing Vision and Other Systems

When it becomes necessary to connect the robot to more complex systems (such as vision), there may be a much more difficult interface requirement. In general, these systems should be interfaced by the manufacturer of the robot and offered as part of a system option.

Some of the newer languages (such as RAIL) have made this interface very simple for the programmer. The user *must*, however, understand the operation and limitations of the vision system. It is a relatively common mistake for systems engineers to greatly overestimate the power of vision systems. This overestimation is especially true with regard to the ability of the vision system to distinguish objects in the presence of less than perfect backgrounds. For this reason, the reader is urged to consult the reference material from the manufacturer (and Chapter 4) before becoming committed to a system requiring vision.

Definition of Special Data Types

Although most high-level languages offer several predefined types of variables (such as real numbers, integers, arrays, etc.), a robotic language requires additional type definitions. These types of variables can include point, path, plane, and frame. Although it is possible to circumvent special data types, they can greatly facilitate programming.

In some languages (such as BASIC), the data types are fixed. In other languages (such as Pascal), new data types can be defined based on the data types already contained in the language.

The preceding features are among the most important for an efficient robotic manipulator language. Different robot languages offer various combinations of these and other features in a variety of different formats. One of the most complete and interesting languages is *RAIL*.

The RAIL Language

RAIL is among the best robot languages at this time for two reasons: First, Automatix Inc.[M7] is a relatively young company, having been formed in January of 1980. This means that the company had no compatibility problem with earlier products. They could start fresh. Second, the company was better financed than most (10.5 megabucks by February 1981). This allowed Automatix Inc. to attract some extremely good talent. The result has been an excellent language that makes a good example to demonstrate the implementation of the features we have been discussing because RAIL has most (though not all) of these features. It should be mentioned that this description of RAIL is included here only as an example of a robot language, not as a tutorial. The description that follows is not a complete one.

RAIL is an extension of Pascal, and it is assumed here that the reader is at least vaguely familiar with Pascal. Readers who have had experience with C, Ada, PL/1, or PL/M should be able to follow these examples fairly easily. A typical hardware configuration running under RAIL is shown in Fig. 2-19.

RAIL offers three special data types: points, paths, and reference frames. These take on the following form:

Points define the position of the tool in terms of the six basic degrees of freedom previously discussed. They are relative to the world reference frame. For example:

START = [100.0, 200,0, 0.0, 45.0, 90.0, 180.0]

would define a tool contact point named START at a position of X = 100, Y = 200, Z = 0. Dimensions are in inches or centimeters (if desired). The rotational orientation of the tool is defined by the last three values (in degrees). These are the angles through which the world reference frame would have to be rotated to align with the tool.

Paths are defined as a sequence of points. For example:

NEWPATH = PATH {POINT1, POINT2, POINT3}

defines a path called NEWPATH that runs from previously defined point POINT1 through POINT2, to POINT3. Points may be any expression that evaluates to a point, including relative positions.

Frames can be defined relative to the world reference frame, either in the program or by calling the operator through the ICM (Fig. 2-20). RAIL allows compound references relative to frames. For example:

MOVE CORNER : [20, 0, 0, 0, 0, 0]

92

Fig. 2-19. A system running under RAIL.

would cause a move to a point that was 20 units (inches or centimeters) along the X axis from the origin of the reference frame CORNER.

RAIL has several special motion commands, including MOVE, WELD, APPROACH, DEPART, ROTATE, OPEN, CLOSE, and CALIB. MOVE can be of the form above, or it can include speed specifiers. RAIL has 11 speed schedules that are set with default values and may

Fig. 2-20. An ICM control pendant.

Courtesy Automatix Inc.

be redefined. A move command with a speed schedule would look like:

MOVE POINT1 WITH SPEEDSCHED[1] AND
POINT 2 WITH SPEEDSCHED[5]

If the straightness of the path is not critical, very fast speeds may be accomplished by specifying SLEW speed:

MOVE SLEW POINT1

Note that if SLEW is used, a SPEEDSCHED command cannot be used. If neither a SLEW or SPEEDSCHED command is present, the system defaults to SPEEDSCHED[0].

WELD commands are similar to MOVE commands, except that welding occurs and that a WELDSCHED for wire feed is specified. For example:

WELD PATH1 WITH SPEEDSCHED[2], WELDSCHED[2]

94

APPROACH causes a move to within a specified distance of the referenced point. The distance is along the negative Z axis of the tool (straight out from the tip), and the orientation of the tool is specified by the point being approached. For example:

APPROACH SLEW 5 START

would cause the tool to move to a point 5 units away from the predefined point START. APPROACH could also use a SPEEDSCHED.

DEPART moves the tool a specified distance away from the point it is at, and in a direction directly opposite of that in which the tool is pointing. For example:

DEPART 5 SPEEDSCHED[3]

would back the tool away from the work (presumably) by a distance of 5 units.

OPEN and *CLOSE* control the simple nonservoed, unsynchronized gripper. Their interface may be reassigned to other than the standard port, if necessary. Because the gripper is not synchronized, these commands have a built-in 0.5-second delay to allow settling of the motion before the next instruction is executed.

CALIB is the RAIL command that *calibrates* its relative motion system to a known reference position (as discussed early in this chapter). It must be executed before any other movement commands.

RAIL allows the operator to define a point at run time using a LEARN command. A reference frame may also be defined using a LEARN FRAME command. The operator is signaled through the ICM which was earlier discussed under relative motion.

Several locations are predefined in RAIL. For example, HERE is equated to the present location. To save the current position for future reference, the statement would simply be:

REFERENCE = HERE

Another predefined location is HOME, which is a set position in which the robot is left in between operations. Likewise TOOL defines the transform of the current tool, and BASE defines the relationship of the desired world reference frame and the base of the robot. All of these features make programming simpler and faster.

The A132 controller (in which RAIL runs) has 16 I/O lines available. These are simple digital, optically isolated signals. The first six are inputs and the remaining ten outputs. Port declarations allow names to be assigned to port pins. For example:

INPUT PORT EMERSTOP 1

OUTPUT PORT READY 13

defines the variable EMERSTOP to port 1 and the variable READY as output 13. When only 1 port bit is equated to a variable, it can take on only the value of 1 or 0. RAIL allows the use of the reserve words ON and OFF interchangeably with these values. More than 1 port bit can be assigned to a variable as

INPUT PORT CONVEYOR _ SPEED 1..4

which assigns port bits 1 though 4 to the variable CONVEYOR _ SPEED.

RAIL allows for communications with the operator through several commands, including READ (read data from terminal), READS (read string from terminal), WRITE (write to terminal), and WRITEICM (write to ICM). The following example illustrates the use of strings. (Courtesy Automatix Inc.[M7])

```
WRITE ('Enter part name')
READS PART _ NAME
IF PART _ NAME < > ('BRACKET') THEN
     BEGIN
          WRITE ('WRONG PART')
     END
WRITE ('Enter average part length')
READ (AVER _ LEN)
WRITE ('Enter part length tolerance')
READ (LEN _ TOL)
PICTURE
IF OBJ _ LENGTH WITHIN LEN _ TOL OF AVER _ LEN
     WRITE ('Part within specs')
ELSE
WRITE ('Part length bad')
```

Notice the ease with which a call to the vision system was made. The command PICTURE caused the vision system to grab and digitize an image, and certain information about the image was automatically equated to reserved variables (OBJ _ LENGTH in this case).

RAIL offers several time-related functions including a WAIT statement, a calendar and time-of-day clock, and a single-interval timer. The interval timer that can be started with the command TIMER ON. Likewise, the timer can be read and stopped by the command TIMER OFF. This timer is a simple hardware interval timer that counts in seconds. The maximum count is 32,767 seconds (approximately 9.1 hours). Although the interval timer is not quite as powerful as the TICK timers of a multitasking executive, it is better than the WAIT function alone.

The time-of-day clock returns the time in string form and is intended for documentational purposes rather than control timing. It could, of course, be used for timing if the proper string comparisons were included with its use.

RAIL contains a fairly complete set of mathematical functions. Besides the basic functions, RAIL includes the conversion functions (TRUNC, FLOAT, and ABS), the geometric functions (SIN, COS, and ATAN), and a square root command (SQRT).

For editing, RAIL includes a modest line editor similar to the ED editor of CP/M. It also features a single-step debugging mode that is very interactive. This mode allows the operator to modify program variables, path points, and speed schedules while stepping backward and forward through the program. The debug mode is well integrated with the ICM pendant.

The following program contains two RAIL functions. The function WELD _ BRACKETS waits for the next part, clamps it down, welds it, unclamps it, and waits for the next part. Every ten parts it calls the second function CLEAN _ TORCH that will clean the nozzle of the torch. (Courtesy Automatix Inc.[M7])

```
;Define ports    (Note: This is a comment in RAIL.)
INPUT PORT PART _ READY 1
OUTPUT PORT CLAMP 3
FUNCTION WELD _ BRACKETS
    GLOBAL BRACKET
    BEGIN
        PARTS = 0
        WHILE CYCLESTOP  = =  OFF DO
            BEGIN
;
;     Move to HOME position, wait for the next bracket
;     then clamp it down and weld it
;
```

```
                    MOVE HOME
                    WAIT UNTIL PART _ READY  == ON
                    CLAMP = ON
                    APPROACH 50 FROM BRACKET
                    WELD BRACKET WITH SPEEDSCHED[2],
                    WELDSCHED[14]
                    DEPART 50
                    CLAMP  == OFF
;
;       Clean torch after ten parts
;
                    PARTS = PARTS + 1
                    IF PARTS = 10 THEN
                        BEGIN
                            CLEAN _ TORCH
                            PARTS = 0
                        END
                    END
                END
OUTPUT PORT BRUSH 1
OUTPUT PORT SPRAY 2
FUNCTION CLEAN _ TORCH
    GLOBAL CLEANER _ BRUSH, CLEANER _ SPRAY
    BEGIN
;
;Brush out the torch nozzle, and then spray it.
;
        APPROACH 100 FROM CLEANER _ BRUSH
        BRUSH = ON
        MOVE CLEANER _ BRUSH
        DEPART 100
        BRUSH = OFF
        MOVE CLEANER _ SPRAY
        SPRAY = ON
        WAIT 1 SEC
        SPRAY = OFF
        DEPART 100
    END
```

Notice that the designation of positions as GLOBAL variables means that they are accessible from anywhere in the program. Before this program could be run, the positions BRACKET, CLEAN-

ER ⏤ BRUSH, and CLEANER ⏤ SPRAY would have to be defined. Defining could be done through the ICM using LEARN statements, or it could be done by specifying the position parameters in the program.

These examples should give the reader a general feeling for the way in which a language can be adapted to robotic control. Many possible enhancements to such a language exist, and a rapid evolution in the years to come can be expected. As more robots appear on the assembly line (Fig. 2-21), problems with synchronization and plant integration will require more elaborate inter-robot communications. Sensory systems will also become more complex. These and other changes will be reflected in the programming languages of the robots.

Courtesy Unimation Inc.

Fig. 2-21. An example of a highly automated process requiring robot synchronization.

SPECIAL PURPOSE MANIPULATORS

Now that we have studied the common types of pick-and-place robots, we should take a brief look at a few more specialized robotic manipulators. In many applications spherical and revolute pick-and-place robots are not adequate. The most common reasons for this incompatibility are

1. Rotary based robots can only approach an item from one direction, and their area of operation is spherical.
2. Even heavy-duty rotary based robots are usually limited to lifting capacities of a few hundred pounds.
3. Additional precision may be required.

One of the most common alternatives to the rotary-based robot is a manipulator mounted on a rail system. Almost every imaginable combination of manipulators and rails has been constructed. At least one system [5] included Unimation Inc. [M6] robots mounted on "air bearing" tracks on the floor. The majority of these systems use overhead rails.

Overhead rail systems vary from extremely precise, light-duty

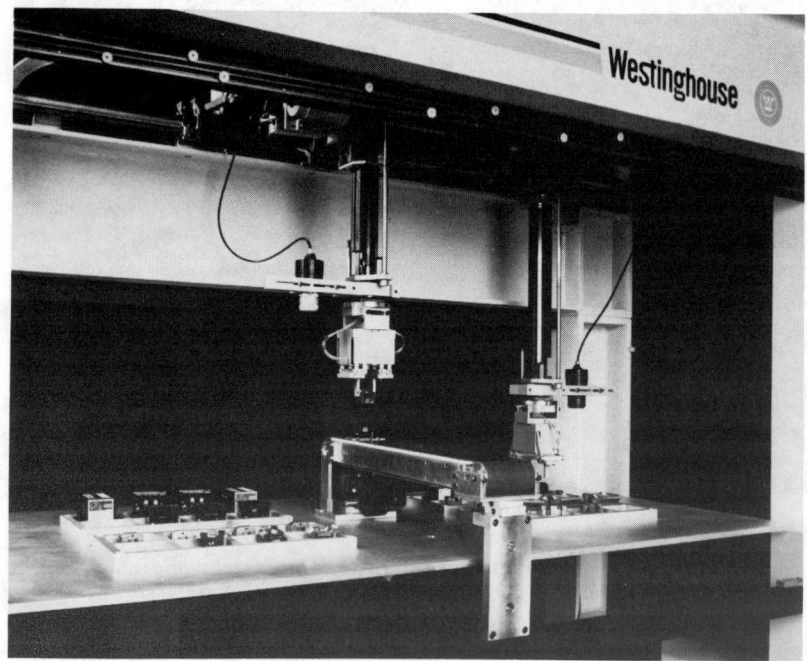

Courtesy Westinghouse Electric Corp.

Fig. 2-22. A two-arm precision assembly robot using overhead rails.

systems to giant cranelike structures. The light-duty, two-armed robot of Fig. 2-22 is typical of the smaller extreme. These light-duty robots find application in the assembly and testing of electronics components, pharmaceuticals, optics, and a variety of other items.

The Mobot robot of Fig. 2-23 has a manipulator mounted on a cross bridge.[M16] This configuration allows the robot to approach an item from almost any direction. Robots of this basic construction are available in very large sizes, and it is not uncommon for such systems to be built on a custom basis to fit an application.

Figs. 2-24 and 2-25 show yet two more configurations. The combination of an overhead and floor rail on the robot of Fig. 2-25 permits very heavy loads to be lifted and moved over a considerable distance.

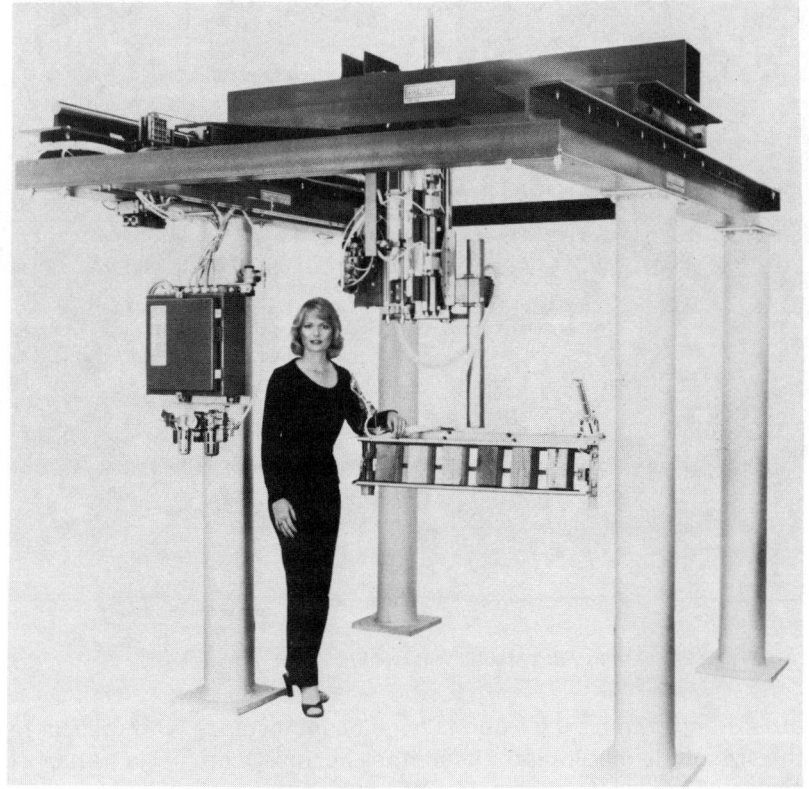

Fig. 2-23. An overhead rail robot with a bridge.

When rails are used in the immediate vicinity of the manipulator, it is natural to consider them as prismatic joints of the robot. On the

Fig. 2-24. A heavy-duty rectilinear welding robot.

other hand, when they extend a significant distance with respect to the size of the manipulator itself, they become more like a transport system. Chapter 3 will consider the subject of robot mobility, why it will become increasingly important, and some of the many ways it is being accomplished.

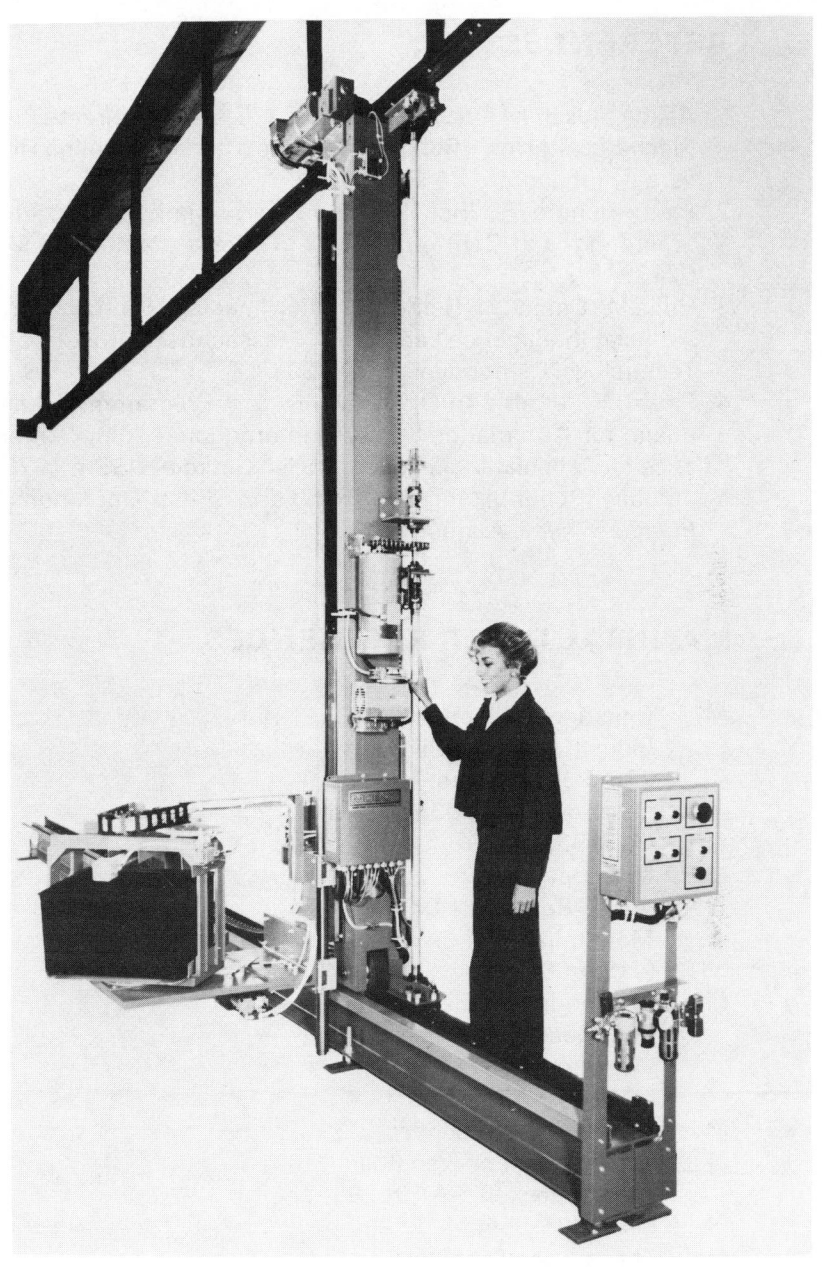

Fig. 2-25. An over-and-under-railed manipulator.

103

REFERENCES

1. Asada, Haruhiko, and Kanade, Takeo. "Design of Direct-Drive Mechanical Arms." Robotics Institute, Carnegie-Mellon University, Pittsburgh, PA 15213
2. Paul, Richard P. "Robot Manipulators, Mathematics, Programming, and Control." The M.I.T. Press, Cambridge MA 02139
3. Hillis, W. Daniel. "A High-Resolution Imaging Touch Sensor." Artificial Intelligence Laboratory, Massachusetts Institute of Technology, Cambridge MA 02139.
4. Finkel, R., et al. "An Overview of AL, A Programming Language for Automation." Fourth International Joint Conference on Artificial Intelligence, Tbilisi, Georgia, USSR, 1975.
5. "Mobile Robot Automates Hot Form Press Operations." *Robotics Today,* August 1982.

MANUFACTURER REFERENCES

M1. Cincinnati Milacron
Industrial Robots Div.
215 S. West Street
Lebanon, OH 45036
(513) 932-4400

M2. Microbot Inc.
453-H Ravendale Drive
Mountain View, CA 94043
(415) 968-8911

M3. Westinghouse Electric Corp.
400 Media Drive
Pittsburgh PA 15205
(412) 778-4347

M4. United States Robotics
1000 Conshohocken Road
Conshohocken, PA 19428
(215) 825-8550

M5. Thermwood Corp.
P.O. Box 436
Dale, IN 47523
(812) 937-4476

M6. Unimation Inc.
Shelter Rock Lane
Danbury, CT 06810
(203) 774-1800

M7. Automatix Inc.
217 Middlesex Turnpike
Burlington, MA 01803
(617) 667-7900

M8. GCA/PAR
3460 Lexington Avenue North
St. Paul, MN 55112
(612) 484-7261

M9. Hodges Robotics International Corp.
3710 North Grand River Avenue
Lansing, MI 48906
(517) 323-7427

M10. Cybotech Corp.
P.O. Box 88514
Indianapolis, IN 46208
(317) 298-5136

M11. Ikegai America Corp.
2246 North Palmer Drive, Suite 108
Shaumburg, IL 60195
(312) 397-3970

M12. ISI Manufacturing Inc.
31915 Groesbeck Highway
Fraser, MI 48206
(313) 294-9500

M13. Mack Corp.
3695 East Industrial Drive
Flagstaff, AZ 86001
(602) 526-1120

M14. Robotic Sciences International Inc.
2709 South Halladay
Santa Ana, CA 92705
(714) 979-6831

M15. Sigma Sales Inc.
6505 Serrano Avenue Suite C
Anaheim Hills, CA 92807
(714) 974-0166

M16. Mobot Corp.
980 Buenos Avenue
San Diego, CA 92110
(714) 275-4300

Chapter 3

Mobility

The first robotic systems to find widespread success in industrial applications were those commonly referred to as *pick-and-place robots*. As discussed in the previous chapter, these devices are in fact simply large manipulators. When compared with the dedicated machines and conveyors that they replaced, these robots appear quite versatile. They can be easily reprogrammed to perform a variety of functions, and they have a far wider field of possible action than the dedicated hardware systems. Which is to say that they are usually associated with a higher degree of control intelligence, and that this intelligence has more manipulative options available to it as a result of the flexible hardware structure.

DISADVANTAGES OF FIXED ROBOTS

Despite the strong advantages of pick-and-place robots, they have some serious limitations. Most of these limitations come as a direct result of their immobility. These disadvantages include:

1. A fixed robotic arm has a limited sphere of operation. The lifting capacity of a robot arm diminishes as its reach is extended. Because of these limitations, pick-and-place robots are most successful in functions that do not involve heavy loads (for example: spray painting, welding, and light assembly).
2. A fixed arm must usually approach an object along a line running from the pivot of the arm to the object (unless it is equipped with an unusual number of joints).
3. If a fixed robot fails, it usually becomes a liability. Since pick-and-place robots are not mobile, they tend to be mounted in locations central to the flow of the product. When a failure

occurs, the disabled robot may block, or partially block, the flow of products. Additionally, the personnel and equipment required to repair or replace the robot may cause further congestion.

4. Changing a plant with a large number of pick-and-place robots from one product or process to a significantly different one may require a considerable expenditure of time and money. Much of this expenditure is the result of the fixed conveyor systems that are typically associated with the robots.

MOBILE ROBOTS IN MANUFACTURING

As a result of these limitations, it is inevitable that mobile robots will find increased popularity in serious manufacturing applications. The combination of pick-and-place robots and mobile robots will be very powerful. In addition to feeding (transporting product to and from) the fixed robots, the mobile robots will be capable of aiding in the retooling and (eventually) the repair of the fixed robots. This combination will require that the systems of the plant be integrated to a large degree under one computer (or a hierarchy of computers). The mobile robots will even be capable of aiding this computer complex in the repair and maintenance of itself and its peripheral systems.

To more fully appreciate the power of such an integrated system, imagine the following scenario. A busy robotic plant is engaged in the manufacture of automotive body parts. The plant is equipped with many of the same machines and fixtures as a plant of today. These fixtures include conveyors, stamping machines, cranes, and pick-and-place robots. All these machines have been fitted with control computers that perform the lower-level functions associated with the machines' routine tasks (as are many such machines in the factories of today). In addition to this primitive intelligence capability, each machine has a communications channel to the main computer of the factory. This link can be used to report production information and malfunctions, and in most cases it can be used to allow the main computer to seize control of the machine.

In addition to these machines, our factory has a number of mobile robots. Some of these robots are simply unmanned forklifts; others are smaller and general purpose in nature. The modular construction of these general-purpose robots allows them to be fitted with special manipulators and other subsystems for performing specialized tasks.

Most of these mobile robots have simple wheeled carriage systems, but one is equipped with leglike structures that allow it access to areas blocked from the other robots by machinery and fixtures.

Suspended from the ceiling of the plant is an array of cameras, each covering a quadrant of the floor. These cameras allow the main computer to verify the position of the various mobile machines.

The activity might go something like this: In one area a truck containing sheet metal is being unloaded by three mobile robots with specially adapted manipulators. These robots are placing the material in storage racks near a group of fixed stamping machines. Two of these robots are on temporary assignment to aid in the unloading of the truck, and the other is part of a two-robot team that normally carries sheets from the racks to the stamping machines. The normal partner of this robot is still feeding the stamping machines as it was doing before the truck arrived. Periodically, one of the robots unloading the truck will bypass the rack and deliver its load straight from the truck to a stamping machine, thus making up for the loss of the robot from that team.

As the last load of sheet metal is removed from the truck, the two robots that were temporarily assigned to the unloading function proceed to a local tooling center where they detach and leave their special manipulators. The third robot rejoins its partner in feeding the stamping machines. One of the two free robots then moves to a replacement center where it will be listed as a ready replacement. The other free robot moves to a second loading dock where it is to assist in loading finished product onto a truck. The instructions radioed to this robot by the main computer were only to proceed to the loading dock tooling center. To prevent cluttering of the main radio (or optional scatter) channel, the program that the robot would need for its new function was not transmitted. Instead, the robot receives this program by ultrahigh data rate, short-range optical transmission, as it approaches the tooling center. Obeying this new program, the robot moves up to be fitted with its special manipulator and then joins another robot in loading the truck.

In another area, the pick-and-place robots are engaged in spray painting stamped parts. These pick-and-place robots are arranged in pairs, one equipped for painting and the other equipped with appropriate manipulators for picking parts off of overhead conveyors (that run the length of the plant), feeding them to its spray painting partner, and placing the finished part on another conveyor. A variety of different parts are traveling on these conveyors in seeming disarray. Some of these parts are headed for the recycling bins at the end of the plant, while others are en route between processes. The main computer is aware of each part and its destination.

The conveyor over one pair of the painter robots is broken and awaiting service. A mobile robot has been temporarily assigned to feed parts from another conveyor to the partner of this painter, thus keeping the painter at 80% of capacity. The stationary feeder robot has been transmitted a program modification to cause it to accept parts from the mobile robot instead of the conveyor.

At this point, one of the other pairs of painter robots reports a malfunctioning spray-painting nozzle. The painter with the malfunction attempts to clear the nozzle by executing a clearing procedure, but the malfunction persists. The main computer is now faced with a management decision. The painter robot needs a new spray-paint attachment, and the mobile robot that would normally get this attachment from the spray paint tooling center has been assigned to load the robot with the malfunctioning conveyor. The nearest spare robot is the one that had been assisting in unloading the sheet metal, but it is some distance away. The main computer calculates that loss of productivity resulting from borrowing the closer robot is less than would be suffered if the robot with the bad nozzle had to wait for the spare robot to arrive. The deciding factor was that the painter with the bad conveyor was only operating at 80% capacity as a result of the makeshift loading procedure. The local robot is thus instructed to proceed to the painting and tooling center as soon as it has given another part to the painter with the conveyor problem. The main computer also considered borrowing the spray nozzle from the painter with the bad conveyor, but it was the wrong size.

At this point, the main computer receives information indicating that the market for these parts is certain to drop significantly and that it is economically preferable to retool the plant to make panels of an alternate design. This change had been scheduled to take place in two days anyhow, and thus it was within the discretionary limits of the main computer to make this change without getting approval from the human management.

As the parts traveling through the system are completed, robots at each station begin to retool for the new products. The spray-painter robots require only a program change, but the stamping machines require heavy new dies to be installed. These dies are too heavy for even the most robust mobile robots in the plant. For this type of problem, the main computer has a heavy-duty overhead crane. The two mobile robots that normally feed the stamping machines are positioned on each side of the stamper that is to be retooled to provide additional video information for the installation of the die by the main computer.

This type of system has a distinct advantage over a plant populated with humans, the presence of a "global" knowledge bank at the main computer. The main computer in such a system would have a *program model* of the entire plant and everything in it. Faced with the decision about which robot should get the spray-painting nozzle, it would simply "think" through each promising alternative by driving its imaginary robots through its imaginary plant. When each alternative was done, it would be evaluated by counting the imaginary products produced by the imaginary machines. All of this would of course take only milliseconds to model.

Another advantage to this global knowledge bank is that the experience of each robot would be immediately available to every other robot through the model. If, for example, a robot encountered unexpectedly poor traction, the main computer might take temporary control of that robot in order to determine the nature of the problem. Assuming that the main computer found the problem to be a grease spill, it would take several steps. First it would add the grease spill to its real-time model, thus no other robot would be instructed to use this route. Next it would check that vicinity of its model for all machines that contained oil. The main computer would interrogate the likely machines for low-oil warnings or changes in performance. It might then use a mobile robot to investigate the likely candidates. Finally, it would run through the possible methods of removing the oil and repairing the leaking machine. The main computer might even be able to predict the flow of the spill, and use this data in its modeling.

It may be true that this scenario goes to an extreme in eliminating human workers. A few human maintenance and managerial personnel will likely be required for some time. On the other hand, it should be remembered that humans will have a negative effect on the order of the factory. For example, no robot would leave a tool in a machine. Additionally, any item that a human might move could cause the global model of the main computer to become inaccurate.

While this whole scenario may seem like science fiction to some readers, it is within the capability of present technology, and all of the aspects are presently in use or under development to one degree or another. They have only to be integrated into one system. The Japanese have already accomplished this integration in at least two factories, although the systems are slightly less sophisticated than the one just described.

Realizing the importance of the mobile robot in this scenario, the Japanese have begun serious development of intelligent mobile machines. Sumatoma Corp. of Japan has even coined a new name for their mobile robot, which is designed to ferry parts. They refer to it as an X-Y Conveyor.

OTHER APPLICATIONS FOR MOBILE ROBOTS

In addition to the functions already mentioned, mobile robots will find application in several other fields. Some of the more significant applications will be mentioned in passing. Notice that each successful niche for a mobile robot (or any robot) will have one or more of the following factors in common.

1. Functions that are dangerous to humans: These functions tend to be associated with high salaries, expensive legal actions, and large insurance costs.
2. Functions that are repetitive, boring, or otherwise unpleasant to humans: These jobs are often plagued by poor quality control, and even outright sabotage. Additionally, the high personnel turnover in these areas often results in abnormally high training costs.
3. Functions that are presently associated with high labor costs: These functions include jobs where high wages are paid for relatively simple tasks, and jobs where a robot is faster and/or more efficient than a human.
4. Functions that require life-support apparatus for human operators: These systems are expensive and they tend to place restraints on the worker. Additionally, the time that a worker can spend at a task is often limited by the air supply of the life-support system. With larger life-support systems, valuable space is taken up by the air, water, food, and sanitation systems.

All of these are basically economic considerations, but they do have a positive effect for the human worker. On the average, jobs open to human workers will be safer, and more diverse, interesting, and creative in nature. The more of these factors that a job contains, the sooner robots will begin to replace human workers in that job. One might think of this as a robotic postulate to Darwin's law.

Mining

Because of the high health risks and labor costs associated with shaft mining, robots will rapidly replace human miners in the years to come. Most of the personnel in a mine today are involved in operating machinery. In fact, today's machines could be thought of as mining robots with humans providing the brain power. The first large-scale mining robots will probably look very much like the present machinery. This leads to an important point that should be mentioned. To find robots operating heavy machinery will be unlikely: They will be the heavy machinery. The skills of the operator will be augmented gradually by intelligent control systems until the machine becomes an independent robot. If a position for an operator remains in such equipment, it will be unoccupied when the machine is running in its normal, automatic mode.

Warehousing Systems

Most manufacturing is done on a large-lot or continuous-flow basis. On the other hand, most orders received by a manufacturer are for relatively small quantities of a variety of different products. An automated warehousing system can serve as an efficient buffer between the stocking and distributing of products.

A relatively large industry exists in automated warehousing equipment, and it has been using mobile robots for many years. Some of these robots look like forklifts, while others look like small trains. Typically, these vehicles follow fixed paths defined by rails or buried wires. The more advanced systems contain computers in each robot, which can retain several operations in memory. A human operator is normally carried by the robot, but the navigation process is usually automatic. The human is presently needed because of the lack of intelligence, dexterity, and vision in these robots. When a "pick" must be made from a bin, the robot signals the human by a display panel or a voice synthesizer.

In warehousing systems where products are all packaged in large boxes or cartons, a rail and pallet system is sometimes used. Small robotic "flat cars" are dispatched down storage sidings where they can pick up or leave pallets of products. These storage sidings are often stacked vertically as well as horizontally. Intelligent cranes or tracked elevators are used to dispatch and recall the robot cars. One end of these sidings is often the input end, and the other is the output.

One approach for picking up and leaving pallets in such a system includes providing the top of the robotic flat car with a small elevator. The pallets in such a system are wider than the cars, and when a car lowers its elevator, the pallet it is carrying comes to rest on a pair of bars on either side (and above) the tracks.

In such systems, a record of the exact location of every item is maintained by a main computer. When an order is received, the computer knows immediately if a sufficient stock of the items required to fill the order is available.

These systems are already approaching the level of integration described in the earlier scenario for manufacturing. As vision systems are improved, there will be less need for human operators in warehousing systems. Completely automated warehousing systems are almost a reality at this time.

Deep-Sea Applications

Because of the stress of high pressure on humans, divers operating below several hundred feet of water must use pressurized suits, or they must ride in submersible vehicles. Once inside the submersible, the operator is in much the same situation as the miner mentioned earlier, that is, entirely dependent on the drive systems and manipulators of the submersible. Since the cost of the intelligence of a robot is usually small with respect to the cost of the "body," it is natural to assume that these machines will also evolve into true robots in time. The limitation here is the degree of intelligence required by the robot. To augment its onboard computer, a submersible can be connected to a surface vessel or platform by armored fiber optic cable. The bandwidth of such cables is high enough to allow the transmission of many channels of video, sensor, and control information.

Space Exploration

Despite the incredible technological success represented by the unmanned spacecraft of the past decade, the National Aeronautics and Space Administration (NASA) is quick to point out that these were not true independent robots. The reason for this distinction is because they were programmed with fixed instructions from earth and could not make significant decisions for themselves. As a result of recent studies,[1] NASA has concluded that a significant part of its research effort in the coming decade should be devoted to the application of robots in space.

Nuclear and Explosive Materials Handling

The handling of radioactive materials is already being done by robotic devices in many areas,[M1] as is the manufacturing of munitions. Most of these devices are not true robots because, like their space-probe cousins, they are under human remote control. Even so, more of the routine functions involving such materials handling will be done by robots in the future, and their functioning will become more and more independent as their computer and vision systems are improved.

Security Robots

Several companies are presently engaged in programs aimed at the development of security robots. These robots will serve to augment and back up conventional fixed-security systems. Since a patrolling robot could be made to pick routes on a pseudo-random basis, it could add an unknown factor to the plans of anyone attempting to breach the security system. Furthermore, since the robot would be in almost continuous radio (or optical) communications with a central computer, destroying the robot would cause an instant system alert. Additionally, the security robot could provide supplementary information to the security computer when readings on conditions in a certain area were ambivalent or contradictory. Since the coverage of a robot can be quite large, it is economically feasible to equip it with more sophisticated instruments than could be deployed on a fixed basis. The mobile security robot will thus be capable of reducing the occurrence of false alarms (a major problem in the security industry).

Agricultural Robots

When watching a harvesting machine operating on the enormous flat farms of the plains states, it does not take much imagination to picture the operator replaced by a computer system. This would already have been accomplished, except that the safety risks were too great. Which is to say that the accidental harvesting of luckless trespassers or a neighbor's dog is totally unacceptable. The ultrasonic scanners and vision systems of the past have been inadequate to do this task. The advent of structured light vision may change this situation in the near future. Such robotic tractor systems could operate as easily in darkness as in light, thus allowing farmers to collect crops more rapidly when frost, rain, or other weather conditions threatened.

The improvement of vision and ranging systems will also allow automated picking machines. Since picking has traditionally been an area of rather low salaries, the introduction of robots may not be as rapid as in other areas.

Military Robots

Much of the funding for the advancement of robot technology, both in the United States and abroad, has military undercurrents. These military applications are especially true of the work being done in robot mobility. Robots can be made impervious to poison gas and low-level radiation and could remain dormant and hidden for extended periods. With an adequate vision system, the warrior-robot could easily react faster than its human counterpart. When captured, a robot would simply self-destruct, thus improving the security of combat information. These and other characteristics could make a robot a formidable warrior and/or spy.

The application of technology to such destructive ends is a sad testimony to the violent nature of people. (On the other hand, I will gladly let a robot take my place in the next war!)

Publicity Robots

In past years, several companies have developed robotlike devices intended for publicity. Some of these devices were used to promote the company's regular products, some were developed to be leased or rented as publicity gimmicks, others have been designed for work in motion pictures, and a few were even intended to deceive unwary investors. Most of these devices have two features in common; they are usually radio controlled by a disguised operator (and thus not true robots), and they are usually designed to have a flashy appearance. Few of these devices actually perform any significant independent task although some do a remarkable job of appearing to be useful.

The robot in Fig. 3-1 is a publicity class robot.[M2] The robot in Fig. 3-1 is called "Sico" by its manufacturer, and it has been widely publicized. The manufacturer of Sico does not pretend that it is autonomous and will candidly admit that an operator controls Sico through the use of switches hidden in his coat. As with most publicity robots, Sico is designed for charm and not functionality.

Fig. 3-1. A publicity class robot.

Despite the fact that publicity robots do not usually perform significant work functions, they can serve at least two positive functions; they can entertain people, and they can serve as goodwill ambassadors between the human population and the approaching army of practical robots. Because of this, publicity robots represent a valid (if rather small) industry.

Domestic Robots

Despite the significant amount of publicity given to the "domestic-servant robots" of the future, domestic work is not a field in which robots are likely to make a significant appearance in the near future. The wide variety of functions required of domestic robots makes them economically unviable (except as status symbols and gimmicks).

The disproportionate publicity given to the domestic robots of the future is reminiscent of the popular theme of the 1950s that showed a helicopter in every driveway. While helicopters have had an impressive impact on our way of life, they are not found in many driveways! As with helicopters, there are (and will be) individuals who build and program domestic robots on an amateur basis, as well as small-scale hobby and novelty robot manufacturers.

Small mobile robots can be used for entertainment around the home, or they can be used for developing software for their larger counterparts. The RB5X robot[M3] shown in Fig. 3-2 is particularly well suited to experimentation. Additionally, some of the robots built by individuals are quite extraordinary, as is Avatar shown in Fig. 3-3.[2] While this activity will help advance technology, it is not likely to make a significant impact on the gloomy economic reality for domestic robots.

Fig. 3-2. Experimental mobile robot RBSX.

Courtesy R.B. Robot Corp.

DESIGNING THE CARRIAGE SYSTEM

Of all the design tasks associated with mobile robots, the design of the carriage system is one of the most important. If the robot cannot get to the work, it certainly cannot perform it. Furthermore, the carriage system is likely to be the major consumer of power in the robot, and thus its efficiency is critical to the performance of the robot. Despite these facts, the task is often approached in a cavalier manner, and poor assumptions are frequently made. Some of the factors that must be considered at the beginning are

1. How critical is efficiency, and how long must it operate before recharging or refueling?
2. Does the robot have to negotiate steps?
3. Does the robot have to negotiate rough terrain or rubble?
4. What is the maximum grade that the robot must climb?
5. Will it be required to operate in mud or snow or on grease?
6. Will the robot be operating on carpets, linoleum, lawns, or other prepared surfaces that it might damage or mar?
7. Does the robot have to carry loads? If so, how heavy and bulky are these loads?
8. What is the minimum width of the doors or other openings that the robot must be able to go through? (Note: Interior doors are usually narrower than exterior doors.)
9. What is the maximum speed and acceleration required of the robot?
10. Can it emit exhaust, and how much noise can be tolerated?
11. Are there weight or dimensional limits imposed by structures or lifts?

Other considerations such as operating temperature range, sun, loading, humidity, and vibration must also be considered, but these are more or less independent of the carriage configuration. For this reason, they were not included in the previous list.

Once a configuration has been found that satisfies the basic criteria listed, there are other serious questions to be asked with respect to its desirability:

1. How complex is the control of the carriage system?
2. What is the minimum turning radius?
3. How easily is the control of the carriage system correlated with the navigation system?
4. Can it be reconfigured (if necessary) for various conditions?
5. Is it able to execute complex maneuvers such as moving in and out of blind corners?
6. How energy efficient is it?
7. What is the reliability, and how much periodic maintenance will be required?
8. How expensive is it to fabricate?

Many of these questions will require quantitative answers if an intelligent choice is to be made. To develop the methods for making

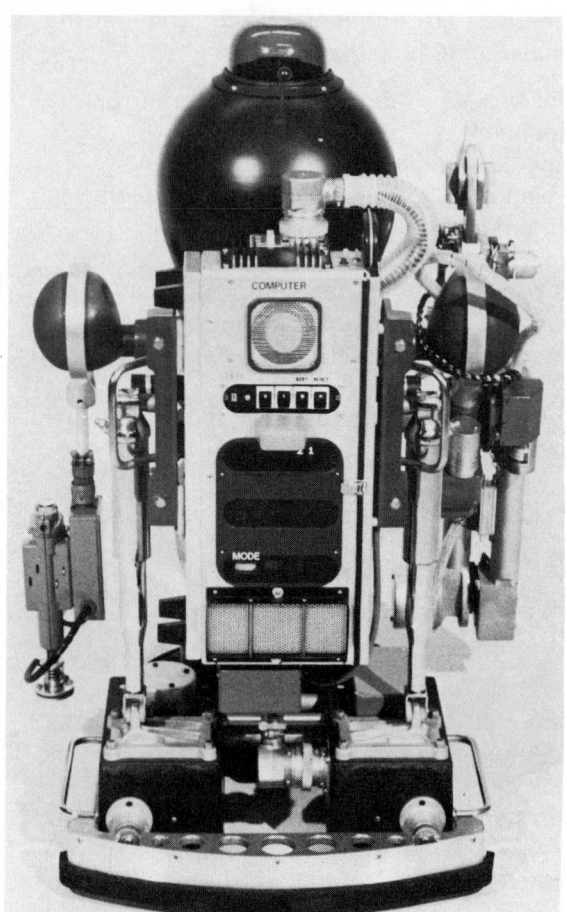

(A) Avatar fully assembled.

(B) Main torso disassembled.

(C) Tractor base.

Fig. 3-3. Avatar and

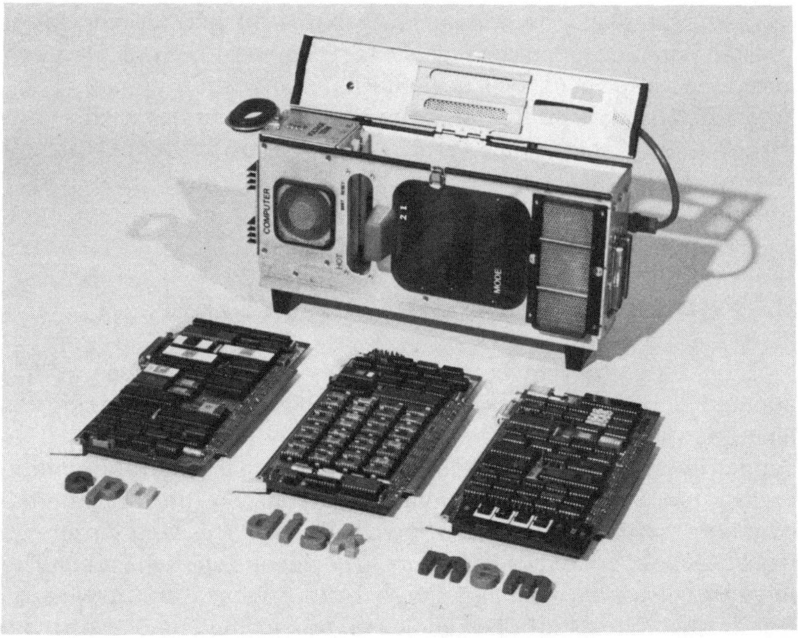

(D) Computer assembly.

(E) Assembled with software development system.

Courtesy Charles Balmer.

its various components.

121

these quantitative evaluations, we will use a simple (and popular) carriage system as an example. Later in this chapter, we will look at several more advanced carriage system designs. Note that although many robotic devices will use internal combustion engines (e.g., automated trucks and tractors), the rest of this chapter will be concerned only with small mobile robots. It will also be assumed that the energy source is a battery although fuel cells and other power sources would be just as appropriate.

THE TRICYCLE CARRIAGE SYSTEM

The basic tricycle carriage system design has been widely adopted for use on early experimental mobile robots. This design has also been popular with hobby and publicity robot designers. This popularity is due primarily to its simplicity of construction and control, and its low cost. There are many variations on the fundamental configuration, one of which is shown in Fig. 3-4. The variation shown uses two independent drive motors to power and steer the robot. The driven wheels are fixed parallel to each other, while the third wheel is free to pivot. Steering is accomplished by causing one of the driven wheels to rotate faster than the other. Relatively tight turns may be accomplished by powering one wheel in a forward direction and the other in a reverse direction.

One of the biggest problems with this configuration is that steering is often erratic due to differences in traction or efficiency between the two driven wheels. This can be partially corrected by placing an angular position encoder on the pivot wheel to help sense the rate of turning. If a pivot encoder is used, some care must be taken during backward movements or differential pivots. Alternatively, the computer can be provided with tight control of the speed of the drive motors. This tight control can be accomplished by connecting a pulse tachometer to each drive motor,[3] or by using synchronous motors such as brushless rare-earth motors and stepping motors.

Another popular variation of the tricycle carriage is to power only the pivot wheel and allow the other two wheels to turn freely. In this variation, steering is accomplished by a gear motor coupled to control the direction of the powered pivot wheel. This alternative is usually less expensive than the first variation, and it allows simpler steering control, but it does not offer very good traction.

Other schemes also use three wheels, but only those designs that have two stationary wheels and one pivoted wheel will be

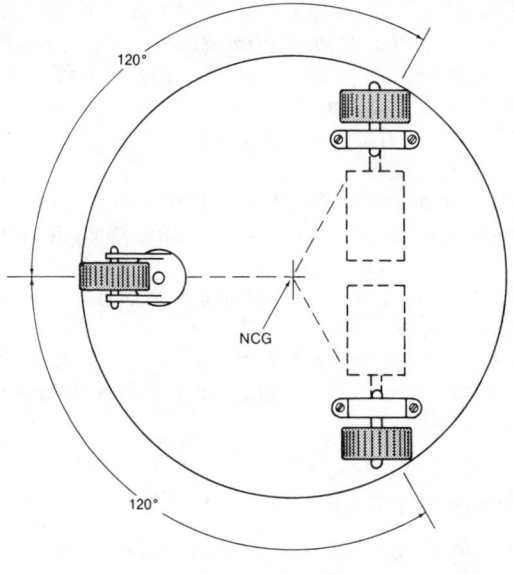

(A) Bottom view.

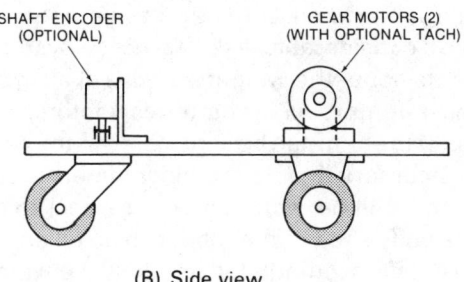

(B) Side view.

Fig. 3-4. Tricycle carriage with rate-of-turn feedback.

referred to as tricycle carriage systems. Some of these other schemes will be discussed later in this chapter.

THE IMPORTANCE OF STABILITY

We will now use the humble tricycle carriage system as an example in analyzing the forces that affect the motion and stability of the robot. The designer who is anxious to get on with the design and programming of the other systems of the robot may be tempted to neglect this task. This is a distinct mistake because one of the most important factors for a mobile robot is its mobility. *A robot that can safely perform its tasks at a higher speed than another robot is proportionally more economical.* Additionally, a robot that can adjust its performance to its load (and surroundings) can be used to safely carry larger loads and thus is more economical than the robot that cannot.

The reason for determining the stability equations is threefold: we want to select the most stable of the competing designs, we need to know the limits of the design we select, and we should include in the program of the robot (and in the program of any host computer) a model of the stability of the robot. The last of these factors is critical since the robot will have to plan its acceleration-deceleration profile to assure that it can stop at the required point while operating at the maximum safe speed.

Determining the Center of Gravity

Before any serious stability calculations can be made, we must determine the approximate location of the center of gravity (CG) of the robot. If the robot is being purchased, the location of the center of gravity should be specified. On a new design, estimating the height of the CG can be a considerable task. The best way to approach this task is to gather data about the weight and size of all of the major components. On most items (such as batteries, motors, and computer systems), the CG of the item can be considered at the center of the item. It will also be necessary to note the height (measured from the CG) at which each part will be mounted with respect to the ground. Each section of the body shell of the robot can be taken as a single object. Once all this data is accumulated, multiply the weight of each object by the distance from the ground to its CG. Adding all these products will give the total vertical moment. Dividing the vertical moment by the total of all the weights will give the approximate height of the CG (Fig. 3-5).

If the results of the stability calculations (that will be discussed shortly) are satisfactory, we will have to assure that the *normal CG (NCG)* falls in the correct place. The NCG is the position on the ground directly below the CG with the robot on a level surface. The method for finding the NCG is the same as for finding the height of the CG, except that both an X and Y moment are found by multiplying the component weights by the distance from their CGs to a reference point. This is a trial-and-error process that can be made much easier by the use of the computer program (CG.BAS) given in Appendix B.

124

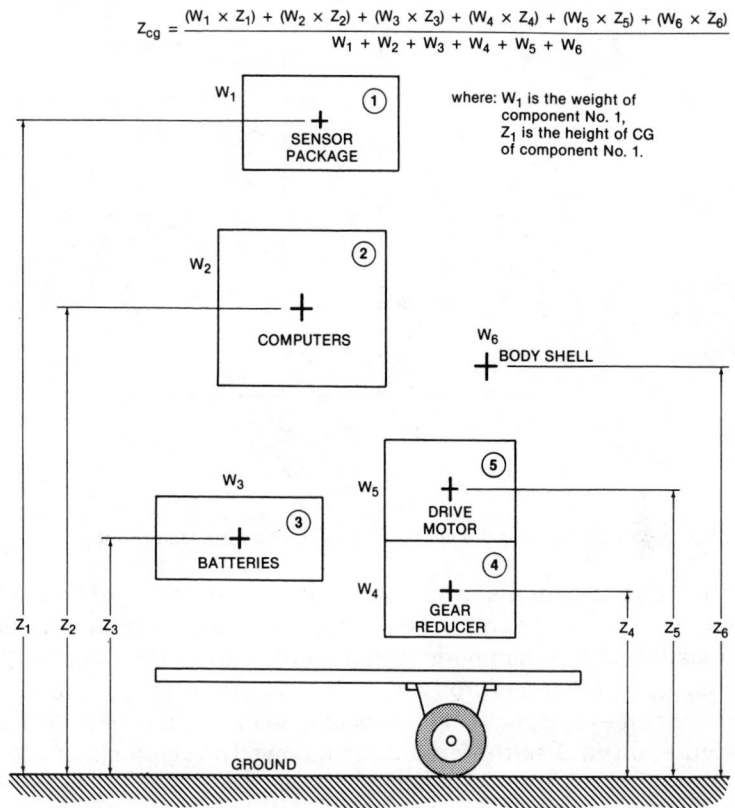

$$Z_{cg} = \frac{(W_1 \times Z_1) + (W_2 \times Z_2) + (W_3 \times Z_3) + (W_4 \times Z_4) + (W_5 \times Z_5) + (W_6 \times Z_6)}{W_1 + W_2 + W_3 + W_4 + W_5 + W_6}$$

where: W_1 is the weight of component No. 1, Z_1 is the height of CG of component No. 1.

Fig. 3-5. Finding the height of the center of gravity (CG) for a robot.

The Program CG.BAS

The CG program (which is a C-BASIC™ program for CP/M®) allows the operator to define and locate components with respect to a coordinate reference system. Fig. 3-6 shows the reference system used in this process. Normally, we are interested in the height of the CG (Z_{cg}) above the ground, and in the displacement of the X and Y components of the CG *with respect to the desired NCG location.* For this reason, the origin of the reference axes has been placed at the desired NCG. Thus, the X and Y values entered in the program are either positive or negative depending on the position with respect to the desired NCG, while the Z values are always positive. As components are interactively added and moved with the CG program, the operator is given a constant update on the location of the CG, the total

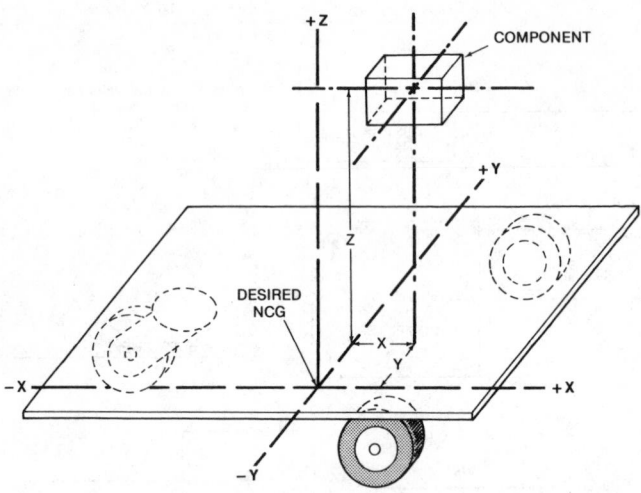

Fig. 3-6. Determining the center of gravity of a proposed robot.

number of components, and the total robot weight. Additionally, when the position of a component is to be entered, the program gives the location of that component that would center the CG over the desired NCG. Under this program, components can be named and the operator can even specify the units of measure to be used. When a configuration has been entered, it can be saved on disk for later use. A sample run of CG.BAS is included in Appendix B with the program, and the adjustment process for trimming the CG to the desired location can easily be followed.

It is also possible to balance the robot physically as it is constructed, but this may cause some serious rework problems if the structure is very complex. Skipping this process during the design of the robot can be a very expensive mistake.

Static Stability

As a matter of convention, this discussion will adopt the terminology used to describe the attitude of aircraft. The term *pitch* will be used to describe the angle of deviation from level along the axis of motion (see Fig.3-7), while the term *roll* will be used to describe the angle of tilt relative to level in a plane normal to the direction of motion. Positive pitch will be defined as a "nose-up" attitude, and positive roll will describe a tilt to the left of the robot.

Assuming that we are attempting to determine the suitability of a

126

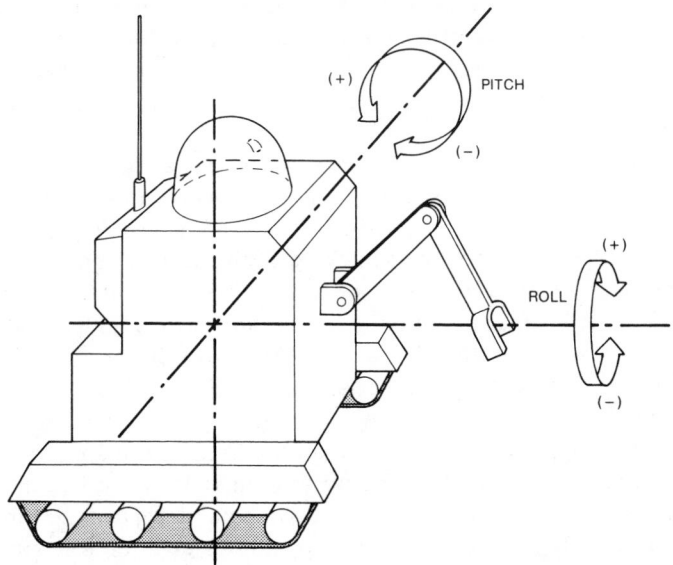

Fig. 3-7. The definition of pitch and roll.

tricycle carriage for our application, one of the first factors of interest is the static stability of the robot. We will assume that it is possible to arrange the heavier components (such as motors, batteries, and gear boxes) in such a way as to place the CG directly over the center of the *zone of stability* as shown in Fig. 3-8. The zone of stability is simply the area bounded by lines drawn between the support points (in this case the wheels).

Once we have determined the zone of stability and the location of the CG, the *critical static pitch angles* can be found by drawing a line from the CG straight forward and backward to the intersection with the edge of the zone of stability (points A, B, and B' on Fig. 3-8A). (Notice that the rear-support point changes according to the rotational position of the rear castor.) The critical static pitch angles can be found graphically as in Fig.3-8B, or they can be calculated as:

$$\theta = \text{ARCTAN} \left(d/Z_{cg} \right)$$

where,

θ is the critical static stability angle,
Z_{cg} is the height of the center of gravity,
d   is the distance (in the direction of interest) from the NCG to the edge of the zone of stability.

127

The critical static positive pitch angle is the maximum incline that the robot could stand on (facing up a ramp) without tipping over backward. Conversely, the critical static negative pitch angle is the maximum slope that the robot could stand on facing downward without falling over forward. The same process could be used to find the critical static roll angles, but notice that there is a worse case for the tricycle carriage. The worst case for this robot is represented by pitch and roll at the same time, caused by a slope in the direction of points E and E' (Fig. 3-8A). If the robot is backing or pivoting on a grade, and the castor wheel is rotated to position B ', the critical stability angle is only 11°! Luckily, robots are usually endowed with considerably more intelligence than other vehicles, and thus can be programmed to avoid dangerous postures such as this.

A robot can drive up or down a ramp very nearly at its critical static pitch angle, as long as it remains at a constant speed. On the other hand, if our robot is driving down a steep ramp and attempts to slow down (or attempts to accelerate while climbing a steep ramp), it may fall over. Because of this problem, our design should be stable at angles considerably worse than those it is actually expected to negotiate.

Dynamic Stability

The force that acts on the robot as a result of longitudinal acceleration (deceleration) is given by the equation:[4]

$$F = M \times A$$

where,

F is the resultant force (in lbs or newtons),
M is the mass of the robot (in slugs or kilograms),
    Note: At sea level, slugs = weight in lbs/32,
A is the acceleration (in ft/sec^2 or meters/sec^2).

This can be shown as a single force pushing in the direction opposite to the acceleration of the robot, and along a line parallel to the acceleration running through the CG of the robot (see Fig. 3-9). In the case of the robot shown, the force is due to deceleration and is acting

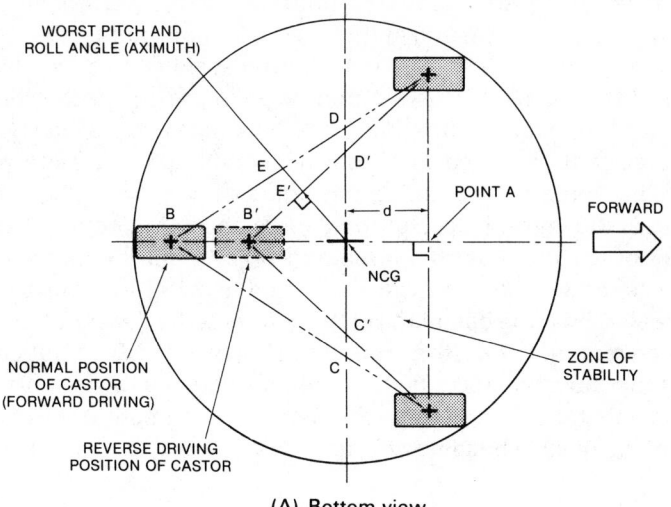

WORST PITCH AND
ROLL ANGLE (AXIMUTH)

D
D'
E
E'
B
B'

d

POINT A

FORWARD

NCG

NORMAL POSITION
OF CASTOR
(FORWARD DRIVING)

REVERSE DRIVING
POSITION OF CASTOR

C'

C

ZONE OF
STABILITY

(A) Bottom view.

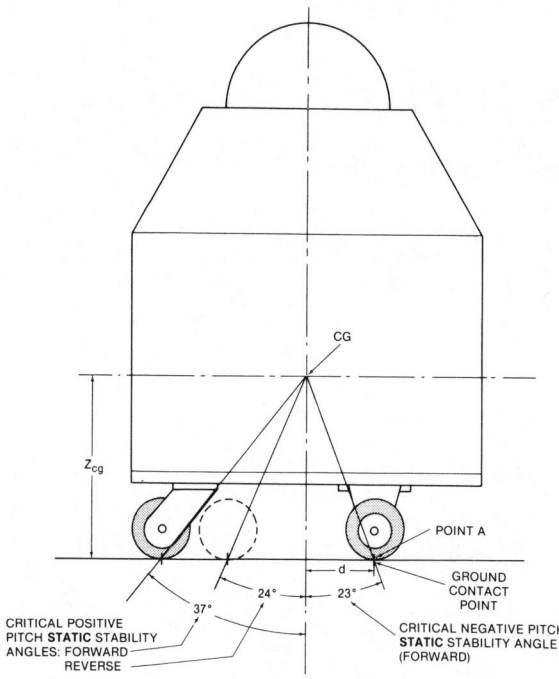

CG

Z_cg

O

O

POINT A

d

GROUND
CONTACT
POINT

CRITICAL POSITIVE
PITCH **STATIC** STABILITY
ANGLES: FORWARD
REVERSE

37°   24°   23°

CRITICAL NEGATIVE PITCH
STATIC STABILITY ANGLE
(FORWARD)

(B) Top view.

Fig. 3-8. Using simple geometry to find the critical angles of static stability.

129

to keep the mass of the robot in motion (we call this *inertia*). Since the breaking action is occurring at the ground contact points, the inertial force can be shown to be a moment acting at a distance equal to the height of the CG (Z_{cg}). This moment is acting to cause the robot to rotate forward over the line (Fig. 3-8A) between its front wheels. The only force that can keep this from happening is the weight of the robot. This force is also acting through a moment arm (D_{cg}). If the product of the inertial (deceleration) force and Z_{cg} is greater than the product of the gravitational force and D_{cg}, the net moment will be clockwise, and our design will fall flat on its face! In order to more graphically show the balancing of these forces, the deceleration force F_d can be rotated to face the ground as shown in Fig. 3-9. Notice also that as the robot begins to tip forward, the distance D_{cg} is shortened, and the imbalance accelerates the rotation. The importance of a low value of Z_{cg} quickly becomes obvious.

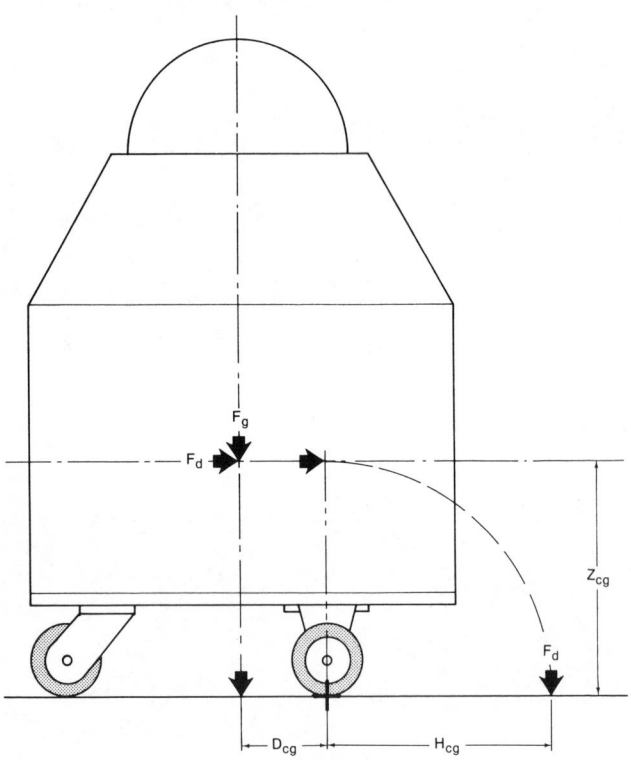

Fig. 3-9. Determining the critical rate of deceleration.

To make the problem more realistic the robot of Fig. 3-10A has been equipped with an arm and is carrying a payload down a ramp. The net moment trying to rotate the robot forward is thus the sum of the static and dynamic moments of the payload and the dynamic moment of the body. If the sum of these moments is greater than the static moment of the body, the robot will flip.

Notice that in Fig. 3-10A, the effect of the slope is taken into account as a modification of the static moment distances. Finding these distances and determining stability is not too difficult for this simple case (See Appendix A, Equation 3-1). However, the method becomes extremely complicated when we consider the more general case of a robot executing a turn on a grade.

An alternate approach to solving this problem is to find the effective CG of the robot and the load. The total effective force acting through this combined CG can then be calculated (Fig. 3-10B). This is done by resolving the gravitational force into a component "normal" to the base (F_{gn}), and a "lateral" component parallel to the base (F_{gl}). In this simple case, the lateral force can then be added to the dynamic (acceleration) force F_d. The effective total force (F_t) and its angle (ϕ_t) can then be found. If a line drawn downward at this angle from the total (combined) CG intersects the ramp (ground) at a point in front of the front wheels, the robot will fall over forward. An expanded version of this technique will be used later in this chapter to determine the general case of robot stability.

Both methods of calculating stability have been shown here because, although the approach of Fig. 3-10B leads to simpler solutions, the first method shows the effects of load position more graphically. While we humans instinctively carry heavy loads close to our bodies, it is less obvious that it is also helpful to carry them low (thus minimizing F_{pd}). The important thing to notice is that the height of the load becomes significant very quickly as the pitch angle exceeds about 20° (Appendix A, Equation 3-1). The height also contributes directly to the dynamic moment. This is the reason that the robot in the upcoming example will go to so much trouble to manipulate its load as it was moving about! It is important to remember that a robot is an *intelligent* machine, and it should be designed and programmed to take advantage of this intelligence. By proper manipulation of a load, a robot can actually improve its stability over its unloaded condition, thus improving its speed and performance! For example, a robot could carry its load behind it (or drive backward) down a ramp in order to improve its stability.

A short trip for a robot is shown in Fig. 3-11. The robot starts at point A and accelerates at full power until it hits a steep ramp at point B. While en route between points A and B, the robot detects that this positive pitch ramp is ahead of it and moves its load out in front of it as much as possible. It also lowers the load as much as possible without running the risk of having the load strike the ramp. As the robot pivots up onto the ramp at point B, it is able to safely lower the load an additional amount, but it must stop accelerating as it is close to instability. Upon reaching point C, the ramp becomes less steep, and the robot is able to accelerate (at less than full power) until reaching

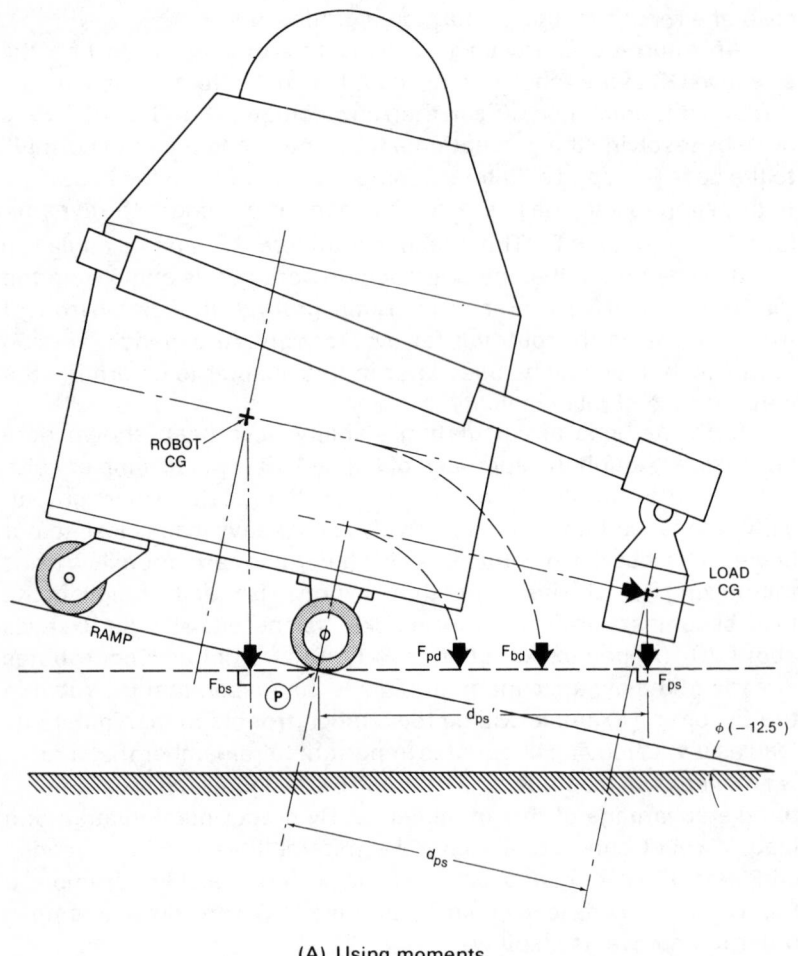

(A) Using moments.

Fig. 3-10. Finding

its maximum velocity at point D, where the grade (coincidentally) ends. Knowing that it must stop by the time it reaches point F, the robot pulls its load in toward its body (to allow for maximum safe deceleration) and begins braking at point E, rolling to a stop precisely at point F.

Dynamic Turning Forces

There is one additional force that our robot must take into account, and that is the centrifugal force due to turning.[4] The robot in Fig. 3-12 is moving at a constant velocity (v) in an arc with a radius (r). Although the magnitude of its velocity is not changing, the direction

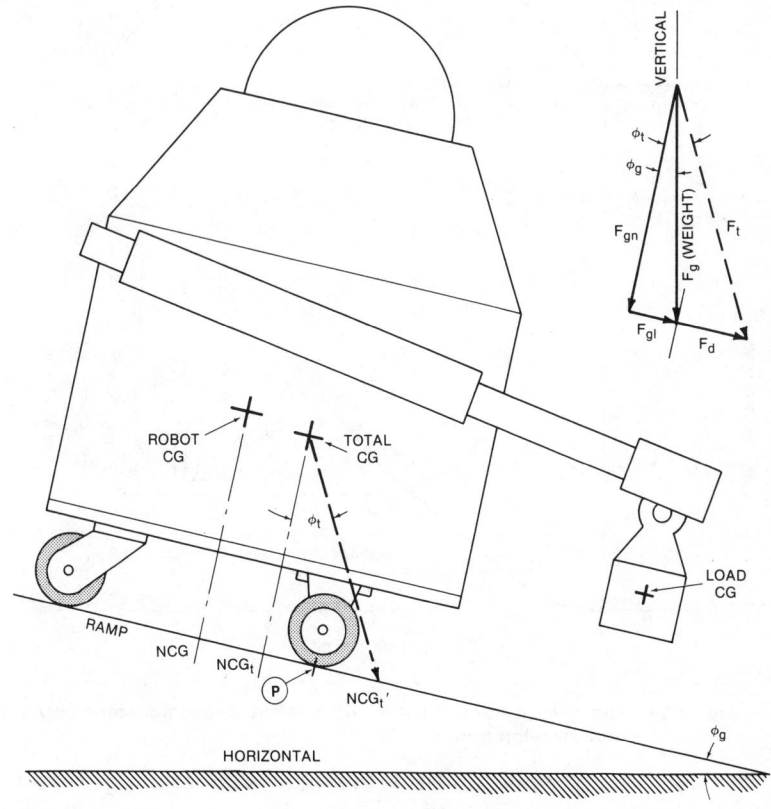

(B) Using vector addition.

dynamic stability on a ramp.

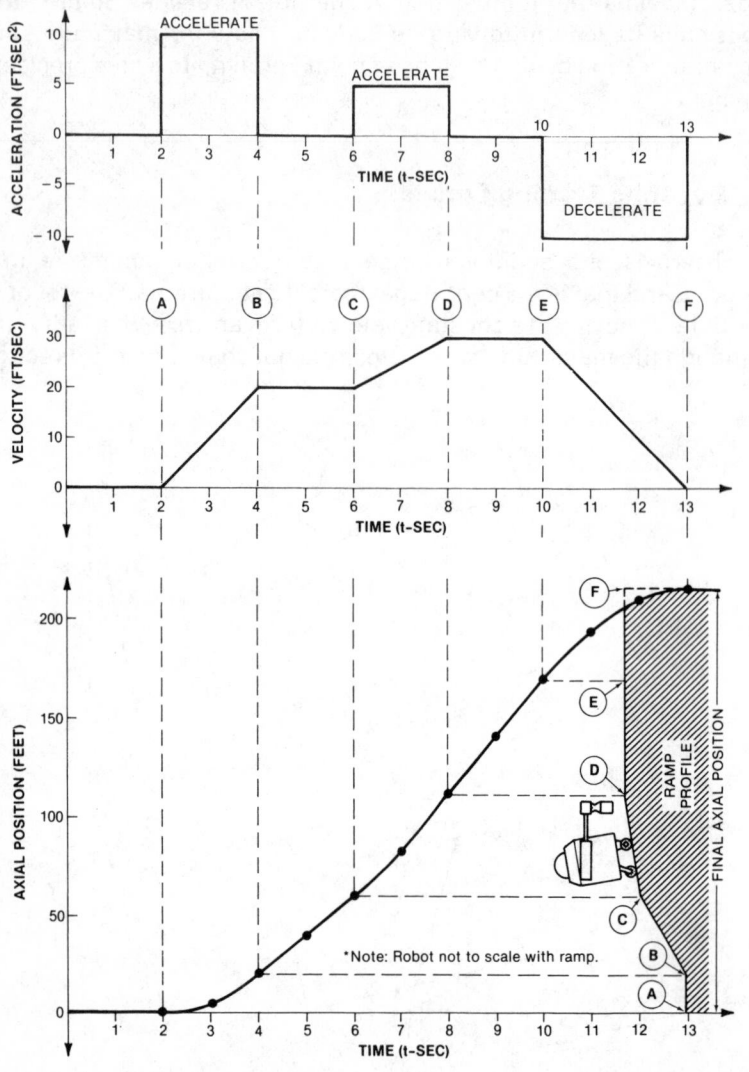

Fig. 3-11. The velocity and position of a robot under constant acceleration/deceleration.

of its velocity is changing, and thus the robot is undergoing acceleration. Each mass of the robot (the arm, payload, and body) will experience a force normal to the direction of motion. (Note that at any given moment, the direction of motion of the robot is a tangent to the

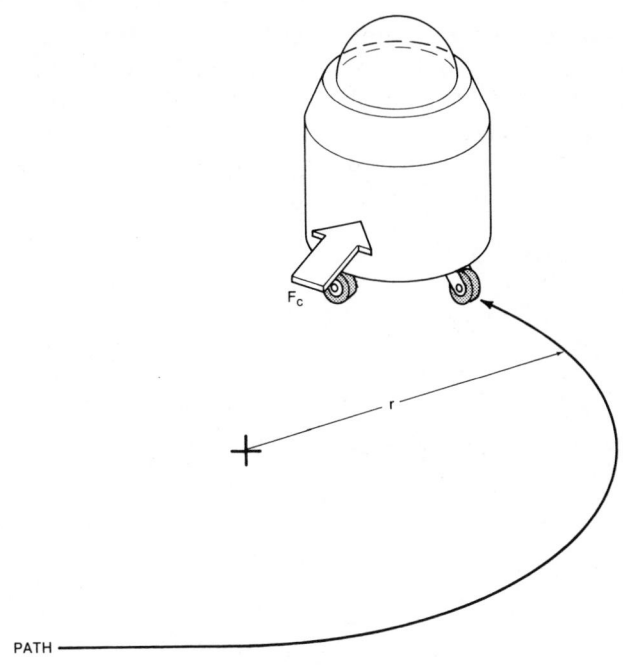

Fig. 3-12. Centrifugal force induced by the robot's turning.

path.) The equivalent acceleration and the net force acting on each mass as a result of this acceleration are given by:

$$A_c = v^2/r$$

$$F_c = (m \times v^2)/r$$

where,
  A_c is the equivalent centrifugal acceleration,
  F_c is the centrifugal force induced by turning,
  m is the mass in slugs (weight in lbs/32) or kilograms,
  v is the longitudinal velocity in ft/sec or meters/sec,
  r is the radius of turn in ft or meters.

In the case of Fig. 3-12, the robot is experiencing a negative roll moment as a result of a turn. No matter how complex the path of a robot may be, its motion at any instant is definable as if it were tracing an arc of some radius (r).

Looking at our robot of Fig. 3-12 from above, we see that as long it does not accelerate, the only force acting on the robot is the centrifugal force F_c. This force is shown as a vector scaled to its magnitude in Fig. 3-13. One can envision this force as deflecting the X and Y components of the point on the ground through which the net force acting through the CG of the robot passes. (This is the three-

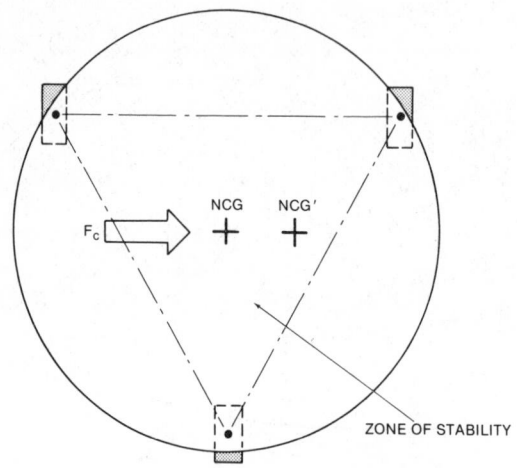

(A) Deflection with turning only.

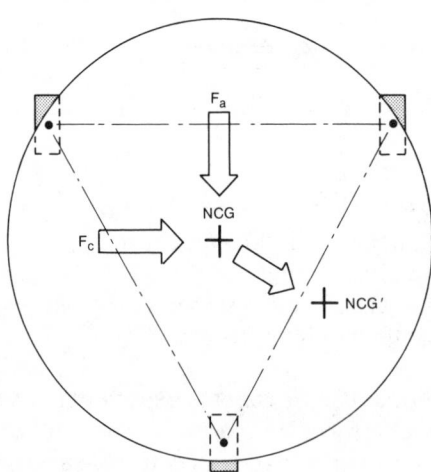

(B) Deflection with turning plus longitudinal acceleration.

Fig. 3-13. Deflection of NCG' on a level surface.

136

dimensional equivalent of the two-dimensional process of Fig. 3-10B.) This position on the ground plane will be designated NCG'. For the constant velocity turn of Fig. 3-13A, the result of this deflection still leaves NCG' within the zone of stability. If the robot begins to accelerate, as in Fig. 3-13B, a new force F_a acts on the robot as described in the earlier discussion. The forces F_a and F_c will produce an equivalent force F_t as shown in Fig. 3-13B. The magnitude of this equivalent force will be the square root of the sum of the squares of the two contributing forces. Notice that this force is in the weak direction (normal to one of the side edges of the zone of stability) discussed earlier in the chapter (see Fig. 3-8A), and will deflect NCG' sufficiently to cause the robot to fall over to its right and rear.

The idea of deflecting NCG' is simply a graphic method of determining whether the component of F_t that is normal to an edge of the zone of stability is sufficiently large to cause an unstable moment about that edge. Obviously, it is not sufficient for the robot to calculate only its pure pitch and roll stability, but it must calculate the net force acting on it (or that would act on it if it executed a maneuver under consideration), and determine if this force will cause an unstable moment about any edge of the zone of stability. With the tricycle base, the negative pitch stability (the moment about the front edge of the zone of stability) is in a direction normal to this edge, but this will not be the case in later configurations (such as the Synchro Drive), so it is important to take the base orientation into consideration. *A reasonable trade-off is to never allow the robot to be subjected to a moment that is greater in magnitude than the magnitude that would cause instability if it were in the worst direction.* This is accomplished by finding the shortest distance from the combined NCG to any edge of the zone of stability. If the magnitude of the deflection of NCG' is greater than this value, the robot will be considered to be unstable. Obviously, this will degrade performance somewhat, but it is a safe trade-off. This approximation will have less impact on systems with more points of support (e.g., wheels), because the zone of stability will become more nearly round in shape.

If the robot's effective CG is displaced from the center of the robot (e.g., as a result of carrying a load), the dynamic forces will cause a moment about the vertical axis of the robot. If this moment is sufficient to overcome the traction of the wheels, the robot will "spin out." If, on the other hand, the centrifugal force is sufficient, the robot may skid sideways. Both of these cases will cause the robot's control and navigation computations to become inaccurate, and the result may be disastrous. When a robot determines that navigation calculations are yielding erratic results, it should immediately ease off on its man-euver (i.e., acceleration, deceleration, or turning radius), and in-crease the margin of safety used in calculating future maneuvers.

Neither the skid nor spin mode of instability will be analyzed in depth here. Although these types of instability are relatively rare in low-speed vehicles, they may be induced by bumps, slippery sur-faces, or pot holes. Instability may also be induced when the robot jumps off the top of a ramp. The robot's operating safety margin should be increased when it approaches such discontinuities.

It should be noted that if a robot is being designed to carry heavy loads, it may be preferable to place the unloaded location of the NCG to the rear of the center of the zone of stability. This placement will "split the difference" of both the spin moment imbalance and the static pitch imbalance, between loaded and unloaded conditions.

Total Robot Stability and the Program STABLE.BAS

The mathematics for the total stability (baring loss of traction) of a robot has been worked out using the NCG' deflection technique, and is available in Appendix A, Equations 3-2A through K. The final equation for the magnitude of the deflection of NCG' (Equation 3-2I) is used as the heart of the computer program STABLE.BAS, which is provided in Appendix B. This program uses successive approxima-tions to find the stability "threshold(s)" of any parameter given the other parameters. It also has the capability of producing stability tables by stepping through various parameter values. Using this program in conjunction with CG.BAS, a design can be tailored to the application and tested against competing designs. This process can all be done without actually constructing anything, and it can then be done again when the subassemblies have been fabricated (to correct any approximation errors present in the earlier run).

The Equations 3-2 (I, J, and K) can also be adapted for use in the robot's internal stability model, and in the model of a central "god-father" computer. Speed considerations will likely require that these programs be implemented in assembly language or in an efficient compiled language (for example: the C language, PL/1, etc.).

Steering and Stability

It is important that the carriage system design be such that the robot has smooth, predictable control over the effective radius of turn.

138

A glance at Equation 3-21 should warn us that the tighter the turn, the more severe this requirement becomes. Erratic turning can easily cause the robot to go unstable.

Sensing the Pitch and Roll Angles

For the robot to determine its stability parameters, we must provide it with a means of determining its pitch and roll angles. The first reaction is to revert to the avionics (aircraft electronics) solution to the problem. A simplified aircraft vertical gyroscope is shown in Fig. 3-14. The two shaft encoders at the pivots of the gyro cage give the robot a reading of the pitch and roll angles. An alternate approach to getting the information front the gyro would be to mount a solid-state laser diode (or focused LED) below the gyro motor and bounce

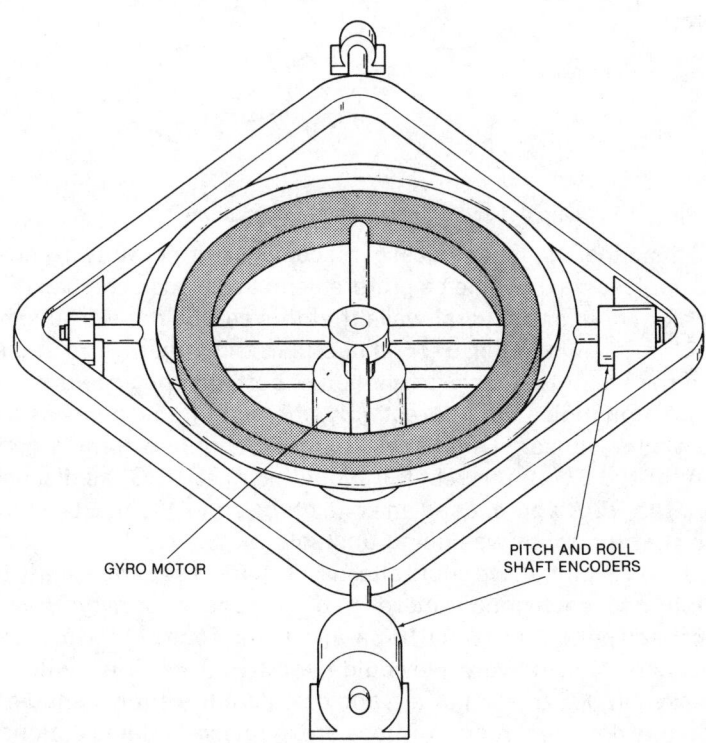

GYRO MOTOR

PITCH AND ROLL
SHAFT ENCODERS

Fig. 3-14. A gyroscope for providing the robot with pitch and roll information.

the beam off of a mirror mounted on the bottom of the motor. The beam would then be reflected onto a beam position detector that would provide signals proportional to the sine of the pitch and roll angles (see Fig. 3-15). In walking robots gyroscopes of some sort will probably be necessary. [5, 6, 7, 8]

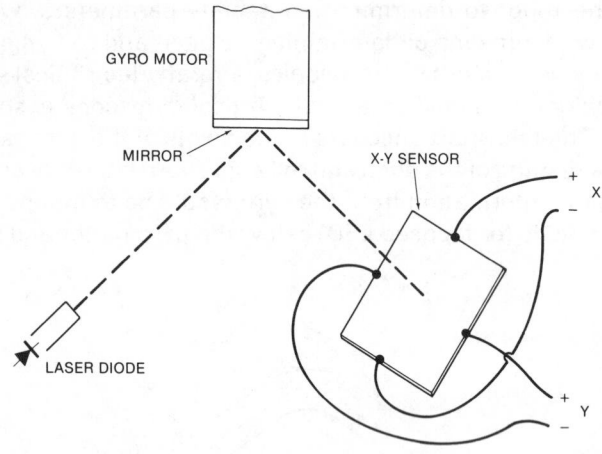

Fig. 3-15. Beam deflection for sensing gyro position.

Unfortunately, mechanical gyroscopes are inherently expensive and troublesome. It would be preferable to avoid such complexities if possible. There are several ways that this can be done. The simple pendulum shown in Fig. 3-16 is useful in this regard. The angle at which the weight hangs is a function of both tilt and acceleration in the direction indicated. Interestingly, the pendulum serves as a net force vector sensor. Thus if we hung a pendulum from a gimbal mount at the CG of the robot, it would point to NCG' as discussed earlier. As NCG' approached an edge of the zone of stability it would mean that the robot was going unstable. Since it is hard to put an encoder on a gimbal, we could alternately hang a separate shaft-type pendulum for each nonparallel edge of the zone of stability. The pivot axis of each pendulum would be parallel to the edge with which it was associated. Alternatively, we could hang two pendulums with their pivot axes at 90° to each other, and calculate the equivalent deflection in any desired direction. These are in fact very viable options for determining the robot's current stability, but they are not sufficient in themselves. The reason for this is that the robot knows only the net force from the pendulum, it does not know what part of that force is

140

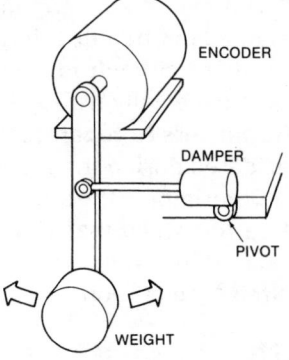

Fig. 3-16. A simple pendulum attitude transducer.

due to gravity (pitch and roll effects), and what part is due to acceleration and centrifugal force. The robot needs this additional information in order to determine the effects it can expect from maneuvers it may be getting ready to execute.

In aircraft, sensors like this pendulum are sometimes used to determine acceleration. This is made possible because the aircraft gyro provides an independent reading of the pitch angle, and thus the effects of pitch can be eliminated from the reading of the pendulum. In the robot we *usually* have the opposite situation; the pitch (or roll) angle is unknown, but we can find the acceleration from a simple tachometer on the drive system. In the case of brushless power motors, the velocity is locked to the rotational field signals generated by the motor control logic, and a tach is not even required. It is true that this assumes that traction is perfect, but significant losses of traction can be detected by other systems on board the robot (such as vision, ranging, and drive motor current monitoring). The mathematics required to determine the pitch and roll angles from the pendulum angle, radius of turn, velocity, and the acceleration are provided in Appendix A, Equation 3-3A and B.

DRIVING THE WHEELED CARRIAGE

Much has been said about the use of conventional dc motors in robotic drive applications. It has long been the consensus that when standard series, shunt, or compound dc (brush-type) motors are to be used, pulse proportional control is the best option. This technique is explained in Chapter 1 and will not be repeated here. It will be pointed out, however, that braking should be done in such a way as to recover as much of the kinetic energy of the robot as possible.

The use of stepping motors and brushless dc motors was also explained in Chapter 1. In the case of these motors, it may be somewhat more difficult to accomplish energy recovery during braking. This is especially true of brushless motors that use internal solid-state sensors to generate the commutation. This trade-off should be considered during the motor selection process.

Determining the size of the motor required is a matter of deciding the performance that will be expected of the robot. The propulsion force required will be determined by the mechanical losses and the desired acceleration:

$$F = (m \times a) + F_d$$

where,

m is the mass of the robot as discussed earlier,

a is the desired acceleration,

F_d is the drag force. (Note that this may be a complex function of speed.)

The key factor in determining the propulsion force is the torque of the motor. To determine the propulsion force of a motor at the ground contact point, we simply use the following equation:

$$F = (T \times K) / r$$

where,

F is the propulsion force,

T is motor torque,

K is the gear reduction ratio,

r is the radius of the drive wheel.

Notice that the torque must be in the same units of linear measure as the radius (r) of the wheel (e.g., inch/lbs and inches). The tricky part of this equation is that the torque of a motor depends on a lot of factors, including the speed of the motor and its internal structure. For the conventional motor types, the typical torque versus speed curves are given in Chapter 1. Relatively precise curves should be available from the manufacturer of a particular motor. The designer will have to decide which basic speed/torque curve fits the application best.

142

Traction

In most industrial applications, traction demands will not be excessively high, but slippage can badly hamper the robot's ability to predict its stability. The two times that this is most important are when the robot is using a drive tachometer to determine velocity and acceleration, and when the slippage causes the robot to miscalculate its radius of turning (see earlier sections).

Some systems (such as treads) offer excellent traction, but at a very high price in other areas of suitability. A common pitfall in robot design is to make the assumption that the more traction a robot has, the better it will perform. This is not necessarily true, as will be seen in the discussion of specific carriage designs.

Climbing, Crawling, and Walking

Many functions will require robots to have carriage systems that are capable of climbing and walking over rough surfaces. Mining and deep-sea applications will probably be among the first to see the practical use of such robots. Robots associated with nuclear and explosive disaster control will also need such carriages.[9] This capability has been demonstrated on several research robots as will be discussed later.

In the minds of many people, a robot must be able to climb stairs, and indeed this is a significant requirement in some applications (such as disaster control). Unfortunately, the need for this capability is often overstressed by enthusiastic proponents. This overemphasis may be a result of the "domestic-robot" fixation mentioned earlier. This can be a serious mistake since these systems tend to be expensive, hard to control, power hungry, and difficult to correlate to the navigation system. In most industrial applications, it would be much less expensive to add ramps, elevators, and/or hoists for the robot population than to buy large numbers of these more expensive robots. Also, few robots will likely have the need to move from one floor level to another in such an environment. Nevertheless, when this capability is required, there may be no substitute for a robot with this type of carriage.

Ease of Control

It is important that the robot's navigation computer be able to interface with the carriage system in a smooth and predictable manner. Some carriage systems (such as legs) offer fantastic stability and versatility, but are a nightmare to control. Ideally, the navigation computer should be required to send only two simple commands to the carriage system: speed and rate of turn. It may be (and usually is) desirable that the carriage system be able to operate in more complex and flexible ways, but the presence of these other modes should not preclude the ability to use the simpler mode.

Most mobile robots will develop a terrain model in their navigation computers. The information in this model will come from the vision system, the ranging systems, short-range sensors, and perhaps even from an external computer. If the robot is to operate efficiently, the navigation computer must be able to easily correlate the responses of the carriage system with the model.

Destructiveness

Some carriage systems are inherently hard on the surfaces that they run on. If the robot is to be acceptable, it must not harm carpets, tear holes in lawns, leave black marks on linoleum floors, or otherwise be obnoxious. Although a robot may originally be designed for an environment where damage to the surface it runs on is not important, it is best to leave options open by avoiding this fault.

Conceptual Pitfalls

There is a strong tendency for us to draw our concepts for robotic carriage systems from three sources:[9] nature, man, and current vehicles. These are valuable sources of inspiration, but we must not blindly imitate them. The systems in nature are made of different materials and were evolved with different goals than robots. On the other hand, most of our vehicles were designed without the capability of internal intelligence, and they were designed for greatly different tasks than the robot will be expected to perform. For these reasons, the robot designer must draw from these systems on a very selective basis.

ADVANCED CARRIAGE SYSTEMS

Now that we have discussed some of the factors that govern the mobile performance of a robot, we can consider some of the more advanced designs that have been built and proposed. Each of these systems represents a trade-off, and none is the sole solution to all problems. With the discussion of each system, we will consider the advantages and disadvantages. You may find aspects of one or more that can be combined with another concept to fit the application you have in mind. The possibilities are almost endless, and this is only a tiny cross section of the promising alternatives. The systematic method of evaluating a design is as important here as the actual systems that will be discussed. We will start with a summary of the tricycle system:

The Tricycle Carriage with Dual Drive

- Efficiency: Excellent
- Simplicity of Control: Excellent
- Traction: Fair
- Maneuverability: Fair
- Navigation: Poor unless feedback is used
- Stability: Poor to fair
- Adaptability: None
- Destructiveness: Good
- Climbing: Poor to bad
- Maintenance: Good to Excellent
- Cost: Low

The Tricycle Carriage with Driven Castor

- Efficiency: Good to excellent
- Simplicity of Control: Excellent
- Traction: Bad
- Maneuverability: Fair
- Navigation: Excellent (but only on surfaces with good traction)
- Stability: Poor to fair
- Adaptability: None
- Destructiveness: Poor to fair depending on whether the control program prevents pivoting the castor without drive motion
- Climbing: Poor to bad
- Maintenance: Good to excellent
- Cost: Low

The Quadrocycle

The three base assemblies of Fig. 3-17 were designed for the hobbyist robot market.[M3] The base in the center is a typical dual-motor tricycle carriage, and the right-most base is a variation of the base in the center that we will call a *quadrocycle.* On this base, a second castor has been added, and the motor-driven wheels have been moved to the center line of the base. By moving these wheels, the stability is improved and the robot is given the capability of pivoting on its own axis.

Unfortunately for the design, three points define a plane. If no suspension is provided with this type of base, one wheel may leave ground contact on rough surfaces. If a power wheel loses ground contact, the robot will lose steering control as well as traction.

- Efficiency: Good to excellent
- Simplicity of Control: Excellent
- Traction: Fair to poor
- Maneuverability: Good
- Navigation: Fair to good
- Stability: Fair
- Adaptability: None
- Destructiveness: Excellent
- Climbing: Poor to bad
- Maintenance: Good to excellent
- Cost: Low

Treads

As mentioned earlier, treads offer the robot excellent traction, and they are particularly useful in mud. If we look at the present use of treads in industry, we find that this is practically the only surface on which they are used. In the case of metal treads, there is a tendency to "chew up" any surface that they operate on.

For some unexplainable reason, we tend to have an immediate attraction to the idea of a robot with rubber treads. This attraction is obvious from the number of cartoon robots that sport treads. It is an impulse that should be reconsidered strongly because treads can be very destructive. This problem grows worse with heavier robots. Fig. 3-17 shows a rubber-tracked robot base assembly (far left). The lack of cleats on the outer tread surface of this track base would probably

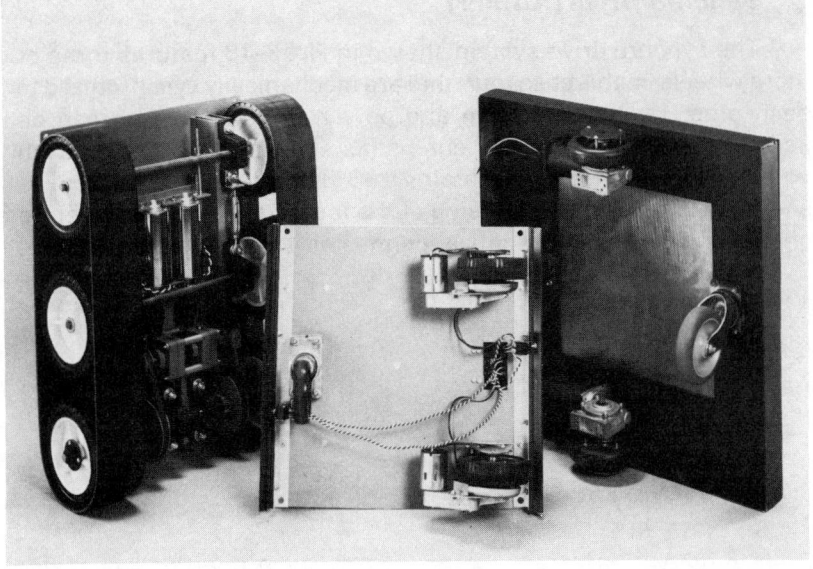

Fig. 3-17. Tread, tricycle, and quadrocycle base units.

minimize its destructiveness, but care would have to be taken to avoid pivots on finished surfaces. This limitation will restrict its maneuverability in many areas.

Some systems have been developed that include small treads mounted on an articulated suspension system.[10, M1] These systems are very useful on soft surfaces, but they tend to be expensive. The most likely uses for these systems will be in space and mining (including deep-sea mining) applications. The evaluation below is for a conventional tread system without articulation.

- Efficiency: Fair to poor depending on the structure
- Simplicity of Control: Good
- Traction: Excellent
- Maneuverability: Fair to poor depending on destructiveness
- Navigation: Good distance, but fair to poor turn coordination
- Stability: Fair
- Adaptability: None
- Destructiveness: Poor to bad
- Climbing: Fair
- Maintenance: Fair
- Cost: Low to moderate

147

Synchro Drive (Author)

The synchro drive system shown in Fig. 3-18 features three or more wheels (in this case four) that are mechanically synchronized to each other for both steering and power. Synchronization can be accomplished by the use of chains (as shown), or by gears. Each wheeled "foot" assembly contains a 90° miter gear arrangement as shown in Fig. 3-19. The housing of the foot is driven by the steering chain, while the inner shaft is connected to the drive chain. The

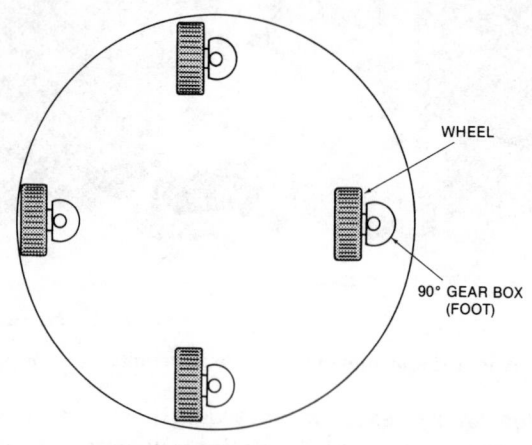

(A) Bottom view.

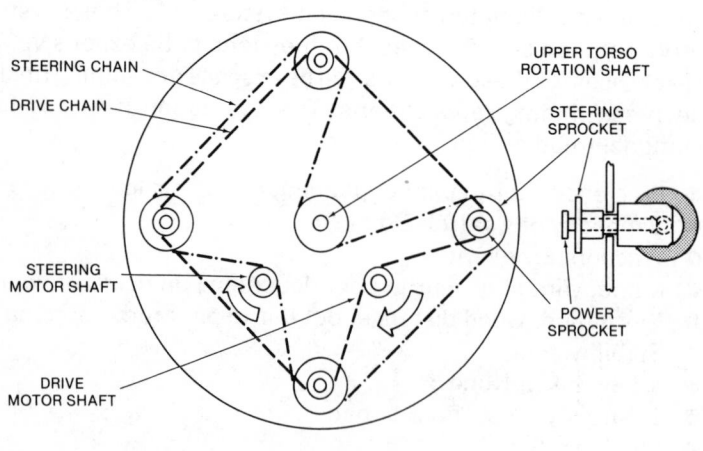

(B) Top view.

Fig. 3-18. Synchro-drive using chain coupling.

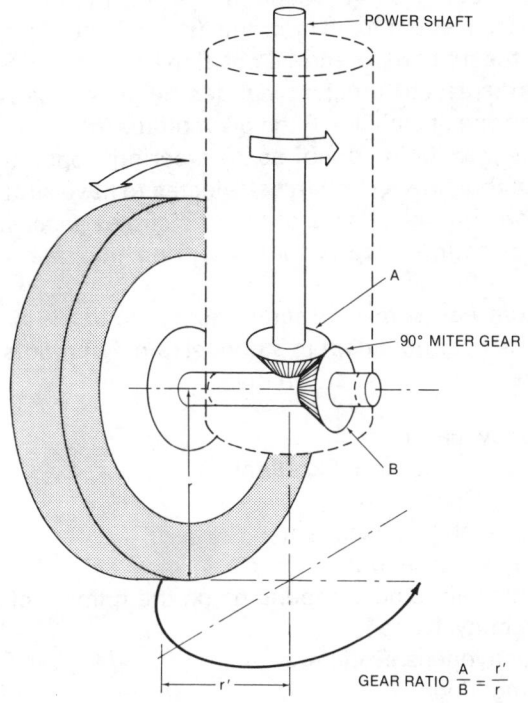

POWER SHAFT

A

90° MITER GEAR

B

GEAR RATIO $\frac{A}{B} = \frac{r'}{r}$

Fig. 3-19. Steering action of the synchro-drive foot assembly.

system offers some interesting characteristics when it is driven. Since the wheels steer together, the base does not change its rotational orientation when the robot executes a turn. For this reason, the upper torso of the robot (which contains the vision and ranging systems) is pivoted and mechanically linked to the steering chain. By driving the steering chain with a stepper motor (and gear reducer) the robot can execute very precisely controlled turns.

Notice that the miter gear (Fig. 3-19) *must* be on the opposite side of the power shaft from the wheel. This is because of the interplay between steering and wheel drive. If the power chain is stationary (the robot is not moving), and the steering chain causes the foot to execute one complete revolution, the wheel power shaft will experience the equivalent effect of one revolution in the opposite direction. With a miter gear having a ratio of 1:1 and located as shown, the wheel axle will revolve once in such a way that the wheel rolls around an arc as shown in the figure. This action is much easier on treated

149

surfaces than having the wheel pivot about its center line. With the 1:1 ratio, the robot will not wobble as the turn is executed, if the wheel radius (r) is equal to pivot radius (r'). Unfortunately, this may mean that in the inboard rotational position (the right-most wheel in Fig. 3-18A) is displaced sufficiently under the robot to cause a deterioration of the zone of stability. If the pivot radius (r') is shortened, the robot will appear to "belly dance" as the steering is operated. If this is objectionable, the miter gear can be selected to have a ratio equal to the ratio of the circumferences of the two circles associated with r and r'. This of course means that it will be the ratio of the two radiuses.

The system has some attractive qualities, but it is not overly stable because it cannot adapt to steep terrain. For robots operating on flat surfaces, it is a good alternative.

- Efficiency: Fair to good
- Simplicity of Control: Excellent
- Traction: Good
- Maneuverability: Excellent
- Navigation: Excellent
- Stability: Fair to poor depending on the number of wheels
- Adaptability: None
- Destructiveness: Excellent
- Climbing: Poor
- Maintenance: Fair to good
- Cost: Low to moderate

Adaptive Synchro Drive with Retractable Leg Assemblies

The adaptive synchro drive represents a compromise between the complexity of a walking robot, and the simplicity (and poor stability) of the synchro drive. The requirement that led to this design was to provide a robot that could negotiate relatively steep ramps, and that could climb over mild curbs without falling over. The solution was to add a degree of adaptability to the synchro drive. This was to be accomplished in the simplest possible manner, and the result is shown in Figs. 3-20, 3-21, and 3-22.

The adaptive synchro drive has the same basic power and steering chain arrangements in the base as did the original system, except that at the positions where the wheels were, there are now pivoted

Fig. 3-20. Synchro-drive leg assembly.

leg assemblies. A set of two chains in each leg transmits the drive and steering control to the same type "foot" assembly used in the synchro drive. Fixed to each leg at the pivot point is a chain sprocket connected to a third stepping motor and gear box arrangement. The power and steering shafts run concentrically through this sprocket into the leg. This arrangement allows the legs to rotate about their pivots (Fig. 3-23), thus changing the effective area of the base. As the legs rotate, the linkages are such that the wheels continue steering in the original direction. Since there is no capability of steering the wheels individually, they cannot be caused to "toe in" during collapsing. For this reason, a certain amount of rolling motion is required to allow this action to take place without damaging the surface on which the robot is running. To accomplish this, the action of the retraction (collapse) motor can be locked to the main drive motor. If the legs must be retracted in a short distance, the robot may have to roll forward and backward once or twice. Alternatively, it can execute a continuous turn with the drive motor turned off. Another disadvantage to this system is that the base orientation cannot be controlled. Thus if the robot approaches a load with the legs extended, it must make do with whatever the orientation of the legs might be. A modification to the steering chain may be added to overcome this problem.

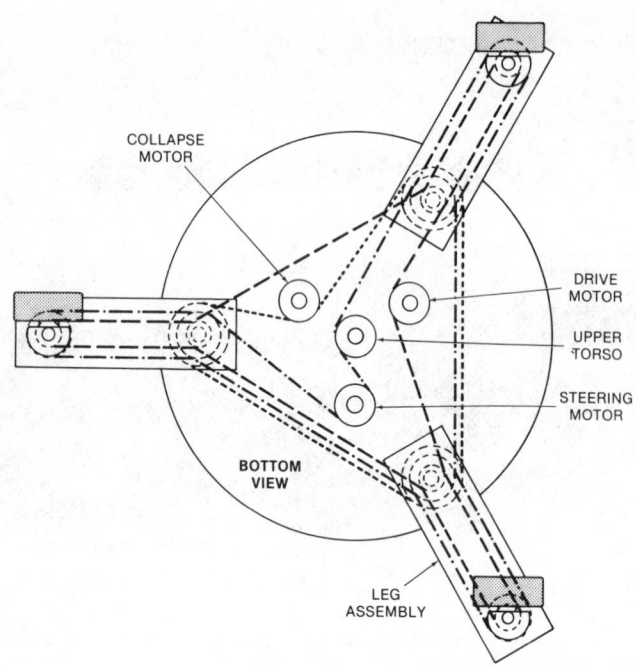

Fig. 3-21. Diagram of a synchro-drive with retractable leg assemblies.

This modification would consist of three movable idlers taking up chain slack between the wheels. In the normal position of these idlers, the wheels would all steer in the same direction. In the other idler position the wheels would all steer tangentially and the robot base could pivot on its own axis. Unfortunately, this reduces the simplicity and economy of the design.

Like the synchro drive, this carriage can steer through 360 degrees without moving. This eliminates the need for backing up. There are several advantages to avoiding this maneuver, including the elimination of rear facing obstacle detection systems and the elimination of polarity reversing circuitry on the main power motor. This second factor improves efficiency since solid-state polarity reversing circuitry always induces some power loss.

This whole arrangement is of course an elaborate compromise, but it was one that satisfied our needs. The lack of independent wheel steering control was traded off for the simplicity of control and economy of having only one drive and one steering motor. The two main sacrifices that had to be made were in the area of efficiency and maintenance.

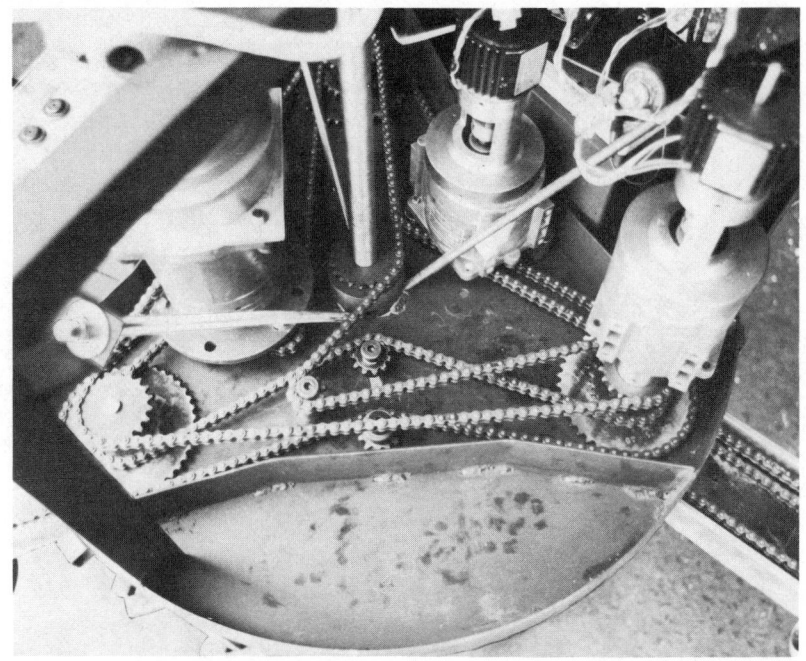

Fig. 3-22. Photograph of a synchro-drive with retractable leg assemblies (battery removed from foreground).

My experimental version of this system (named "Kludge") has been very successful, and the efficiency is better than expected. With 8-inch diameter tires and its legs extended, Kludge can climb over 4 × 4-inch timbers (actually 3.5 × 3.5 inches). My totally unbiased assessment of this approach is:

● Efficiency: Fair to good
● Simplicity of Control: Excellent
● Traction: Good
● Maneuverability: Excellent
● Navigation: Excellent
● Stability: Fair with legs retracted, excellent with them extended
● Adaptability: Adaptable to ramps and small single steps
● Destructiveness: Excellent with precautions mentioned above
● Climbing: Poor
● Maintenance: Fair (I hope)
● Cost: Moderate

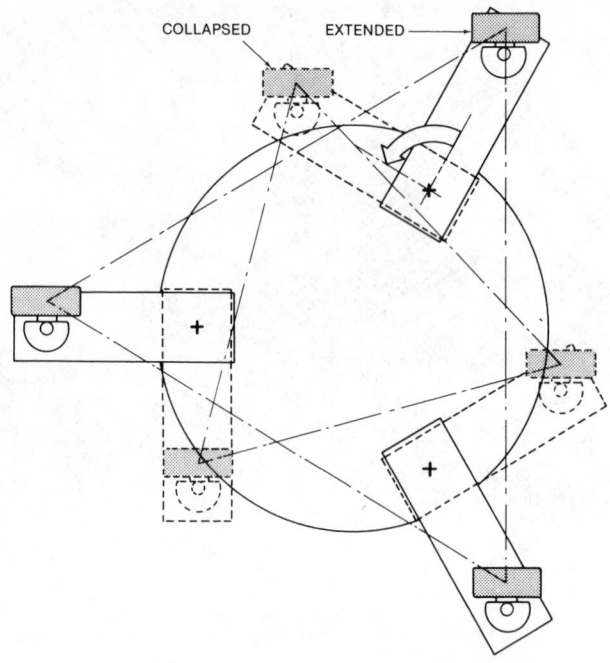

Fig. 3-23. Rotational collapsing and the change of the zone of stability.

The CMU Rover Fully Independent Drive

At the time of this writing Dr. Hans Moravec at Carnegie-Mellon University (see vision) is constructing a new mobile robotic testbed. This configuration (Figs. 3-24 and 3-25) has a similar base arrangement to the synchro drive system, except that each foot (Fig. 3-26) is totally independent for both steering and power. The "feet" also have a wheel on each side, powered through a differential. This arrangement greatly improves the flexibility of the carriage system. Each "foot" assembly contains an encoder for both steering and power, allowing the drive computer to precisely control it. Early results indicate that the differential *must* have limited slip.

The CMU Rover System has joyous degrees of freedom for a wheeled system. It can turn in a mode like the synchro drive, or it can steer its wheels tangentially to the base and spin in circles. The lack of mechanical linkages should provide excellent efficiency, and maintenance should be relatively good. Control is of course more complex,

154

Fig. 3-24. CMU Rover (showing camera slide).

Courtesy Dr. Hans P. Moravec, Carnegie-Mellon University.

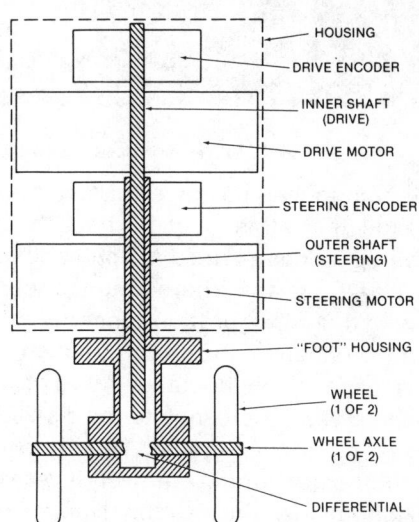

Fig. 3-25. The CMU rover wheel drive assembly (simplified cross section).

Courtesy Dr. Hans P. Moravec, Carnegie-Mellon University.

but the CMU machine is being fitted with special low-power-consumption computers utilizing the extremely powerful MC68000, 16-bit microprocessor.

Since the Rover does not have a configurable base of support, its stability is limited. The combination of these wheels with such a base

155

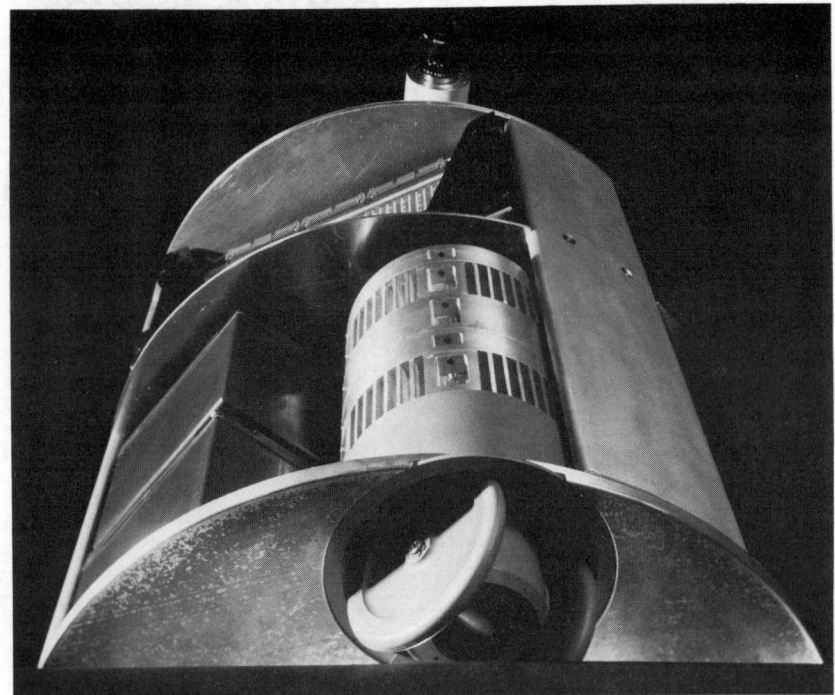

Fig. 3-26. CMU Rover base assembly (showing wheels).

would indeed be powerful, but it was not required for the functions expected of this robot. Perhaps the first law of robotics *design* should be "If it doesn't need it, don't add it!"

One of the interesting maneuvers that Moravec is exploring[11] with this configuration concerns climbing slopes that would normally be too steep for the robot's traction and power (not stability). The robot is made to spin about its own axis as it approaches the ramp. This is somewhat analogous to the zig-zag maneuver performed by a bicyclist on a steep grade. When one wheel is attacking the full grade, another is running relatively level, while the third is moving downhill. Additionally, the rotational motion of the robot serves to store energy in the same manner as a flywheel. While it seems unlikely that one would want to have factory robots pirouetting wildly about the floor, a subdued version of the maneuver might prove very useful.

This system is in keeping with the general trend away from mechanical complexity in electromechanical devices such as printers and manipulators. The electronic controls associated with these devices are being used to eliminate most of the mechanical intricacies, thus reducing cost and maintenance.

- Efficiency: Excellent
- Simplicity of Control: Fair (due to exceptional flexibility)
- Traction: Good
- Maneuverability: Excellent
- Navigation: Excellent
- Stability: Fair
- Adaptability: Dynamic adaptability only
- Destructiveness: Excellent
- Climbing: Poor
- Maintenance: Good

The Omni Drive

Perhaps one of the most esoteric of the wheeled drive systems is the "wheel-within-a-wheel" structure. It has been proposed[12] that Ezekiel may have observed this type of carriage system on an extra-terrestrial craft. If this is so, he did not have much more luck in finding a way to describe it than I have had! A very crude implementation of this concept is shown in Fig. 3-27. The assembly is capable of driving in any direction without changing its directional orientation. In the example, six wheels are supported in a circle so that the outer arcs of their circumferences form a crude composite wheel. The composite

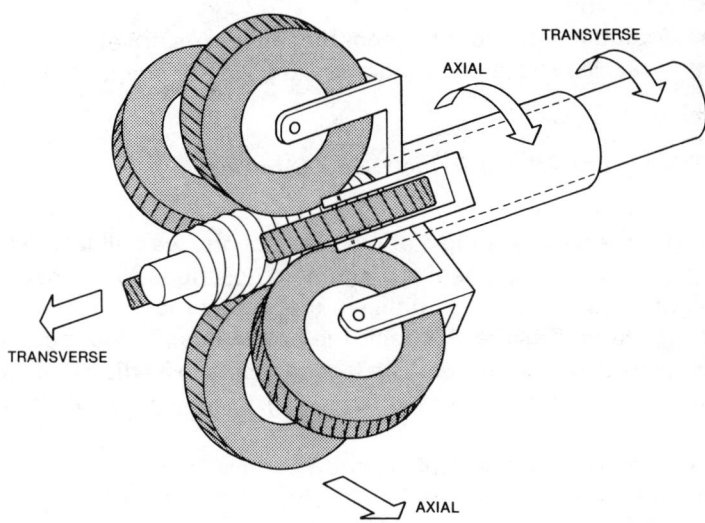

Fig. 3-27. A crude implementation of the "wheel-within-a-wheel" type of drive.

wheel may be driven in the conventional way by rotating the outer drive shaft, while the smaller wheels may be driven by a concentric inner-drive shaft. The result is the omnidirectional motion for which the system has sometimes been named. The main weakness of this system is its complexity.

A version of this drive was developed by the Veterans Administration[13] as a transport system for paraplegic persons (Fig. 3-28). This system powers only the axial motion of each wheel, allowing the smaller outer wheels to roll freely. The effect is that these rollers act as force translators. This effect can be seen for the case of forward drive shown in Fig. 3-29B. Notice that only the two rear wheels are powered for this maneuver and that their outward force vectors cancel. This scheme greatly reduces the complexity of the carriage, but some loss of traction will occur. The small diameter of the rollers will also cause difficulty on surfaces that are not perfectly smooth.

- Efficiency: Good
- Simplicity of Control: Excellent
- Traction: Good to fair (for single-axis drive)
- Maneuverability: Excellent
- Navigation: Excellent to fair (for single-axis drive)
- Stability: Fair to good depending on mounting
- Adaptability: None
- Destructiveness: Excellent
- Climbing: Poor
- Maintenance: Poor to good (for single-axis drive)
- Cost: Moderate (in production) to good (for single-axis drive)

Triangular Wheel Drive (Step Climber)

Before leaving the topic of wheeled robots, we will take a brief look at an interesting wheel arrangement described by Weinstein.[14] As mentioned earlier, it is unlikely that many robots will be seen climbing stairs. Despite this fact, the problem is no doubt going to continue to attract attention. The legged robots to be discussed later will have this capability (and much more), but the price will be very high.

A single-wheel assembly is shown in Fig. 3-30, and four such assemblies would be required on a robot. The mechanism operates with two wheels in contact with the ground while the robot is running on a flat surface. As the forward wheel strikes the front edge of a step, the assembly rotates (hopefully) so that the upper wheel rests on the top of the step. While Weinstein did not claim to have proven the mechanism to operate either mathematically or experimentally, it seems reasonable, and he did work out some of the geometric requirements. The method of steering such a robot was not made

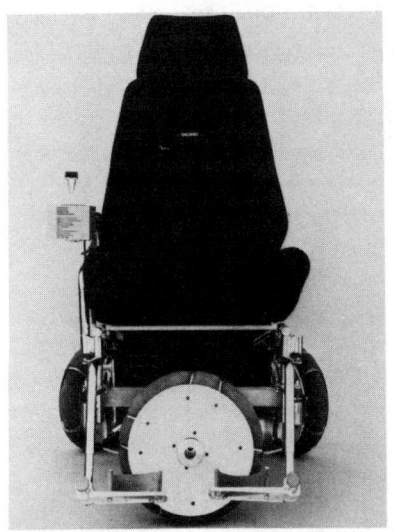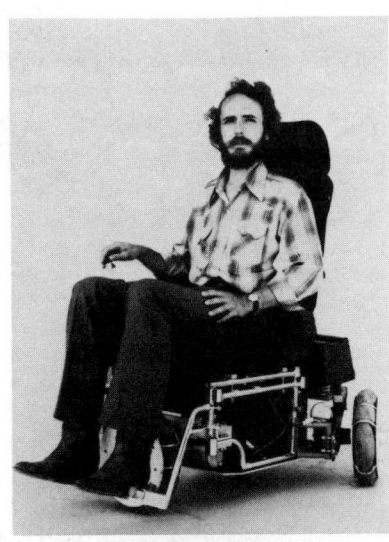

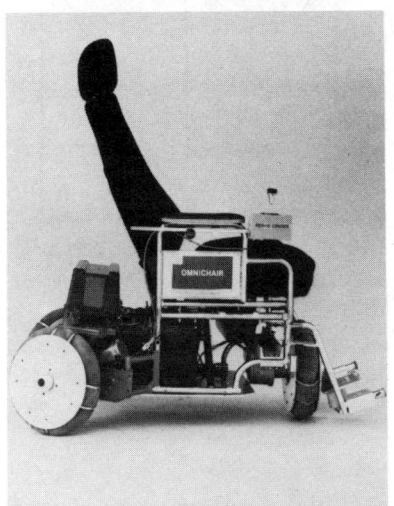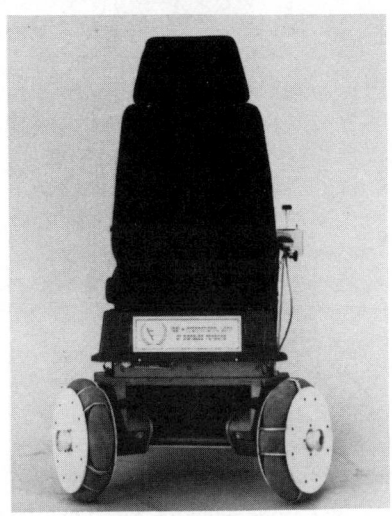

Courtesy Stanford University and Palo Alto VA Medical Center.

Fig. 3-28. The Omni drive.

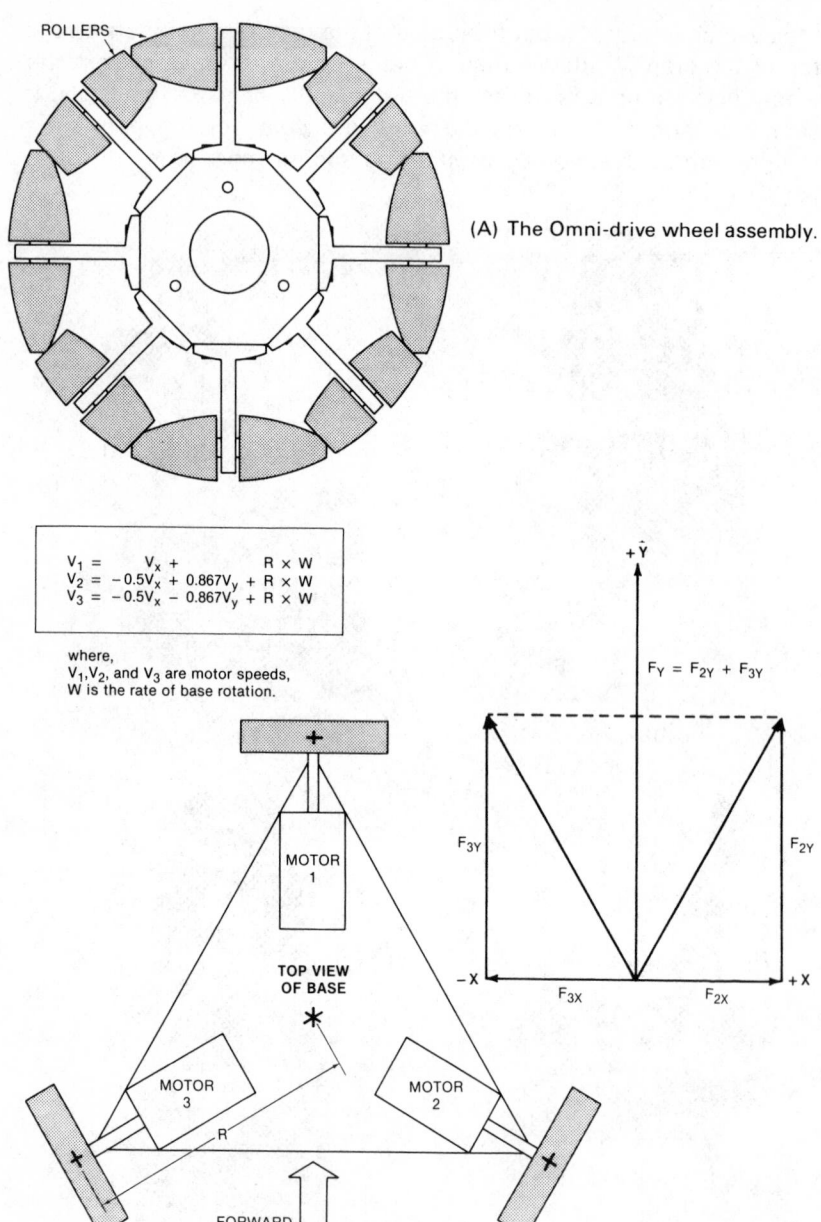

ROLLERS

(A) The Omni-drive wheel assembly.

$$V_1 = V_x + R \times W$$
$$V_2 = -0.5V_x + 0.867V_y + R \times W$$
$$V_3 = -0.5V_x - 0.867V_y + R \times W$$

where,
$V_1, V_2,$ and V_3 are motor speeds,
W is the rate of base rotation.

$+\hat{Y}$

$F_Y = F_{2Y} + F_{3Y}$

MOTOR 1

TOP VIEW OF BASE

*

F_{3Y}

F_{2Y}

MOTOR 3

MOTOR 2

R

$-X$

F_{3X}

F_{2X}

$+X$

FORWARD

(B) The orientation of wheels and force vectors for forward drive (motors 1 and 3 are driven).

Fig. 3-29. Details of the Omni-drive system.

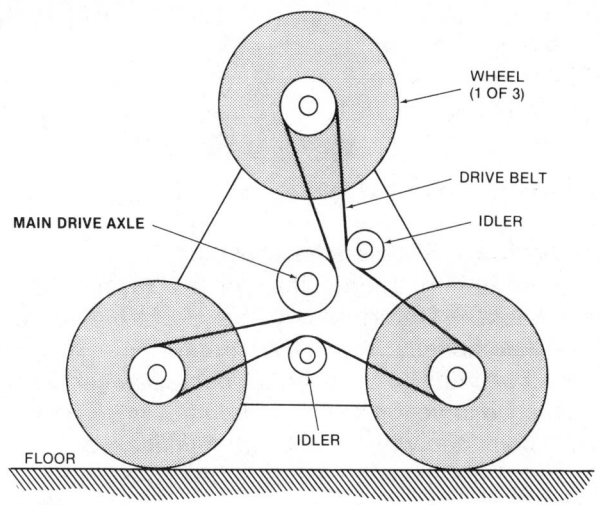

Fig. 3-30. The triangle wheel step climber.

clear, but it could operate in at least two modes. The front set of wheels could be steered like an automobile with the rear set on a fixed axle, or all four sets of wheels could be fixed and it could be steered like a treaded robot. The first alternative would be somewhat destructive in steering (due to the two wheels of each set in ground contact), and the second alternative would be considerably worse.

The success of such an arrangement would seem more likely if the pivoting and climbing action were under intelligent control, rather than functioning as mechanical reflex action. This would also allow for slight trimming of the robot's pitch and roll on sloped surfaces. The destructiveness of this system could also be reduced in this way, but it would, of course, be more complex.

The drive is also restricted to steps of a limited rise and tread (width) and is of little use on rough and broken ground. Even so, it is one of the few simple systems proposed for step climbing.

- Efficiency: Good
- Simplicity of Control: Fair
- Traction: Excellent
- Maneuverability: Fair to poor (depending on steering)
- Navigation: Fair
- Stability: Fair
- Adaptability: Some pitch and roll correction adaptability (if intelligent control is used for rotating assembly)

- Destructiveness: Fair to bad depending on steering
- Climbing: Steps only
- Maintenance: Good
- Cost: Modest

The Hybrid Spider Drive

The robot in Fig. 3-31 is a model of the drive that was considered on the project mentioned earlier (for which the adaptive synchro drive was finally chosen). Unfortunately, the cost of hardware alone was prohibitive. Additionally, the software development time required before the robot could do even a reasonable stagger was discouraging. The synchro drive, on the other hand, was so simple to control that it could be tested with no onboard intelligence.

One of the possible approaches for the construction of the leg is shown in Fig. 3-32. The term *hybrid* has been used to indicate that the system has elements of both a legged and wheeled system. Since the act of walking is relatively slow and very inefficient, this design is capable of operating in either mode to take advantage of the surface it is on. The motorized wheel on the "foot" must be kept small and light in order to minimize inertia during walking operations. For this reason, a trade-off must be made between the use of a fixed powered wheel and an assembly with individual steering control such as that used on the CMU Rover.

With a fixed foot, steering in the rolling mode could be accomplished by using the front legs (and/or back legs) like the steering mechanism of an automobile. The system could also resort to crab like motions in tight maneuvers, or it could operate much like a four-legged roller skater by operating in both modes simultaneously. The more complex the technique, the more difficult will be the control and navigation functions.

With a rotating foot assembly, the Hybrid Spider could perform almost any maneuver imaginable. This could also eliminate the slight destructiveness that would otherwise result from the twisting action of the foot as the robot walks. The high level of complexity of a system like the Hybrid Spider mades for even more tough trade-off decisions than in other systems.

- Efficiency: Walking: poor/rolling: excellent
- Simplicity of Control: Poor/fair to good
- Traction: Excellent/good
- Maneuverability: Excellent/fair to good

Fig. 3-31. The hybrid spider drive.

Courtesy W.G. Spiegel II

- Navigation: Poor/fair to good
- Stability: Excellent/excellent
- Adaptability: Excellent
- Destructiveness: Fair to excellent/excellent
- Climbing: Excellent
- Maintenance: Poor to bad
- Cost: Very high

The OSU Hexapod

At Ohio State University, Robert McGhee and his associates have spent a great deal of effort in perfecting a six-legged insectlike drive system called the "Hexapod" (see Fig. 3-33). This system is not the first of its kind to be constructed. That distinction probably belongs to the General Electric "Quadruped Transporter," which was a manually operated six-legged walker developed in the late 1960s. In fact, McGhee himself developed a four-legged walker at the University of Southern California in 1967. The OSU Hexapod is, however, far more sophisticated in its control than any of these earlier projects. At this time its only rival exists in the Soviet Union,[15] but a similar machine is under construction in Japan.

The Hexapod is being used to develop and test control hardware and software. The system is not intended to be independent, and so it is driven by ac line power through the use of triac controls. The

163

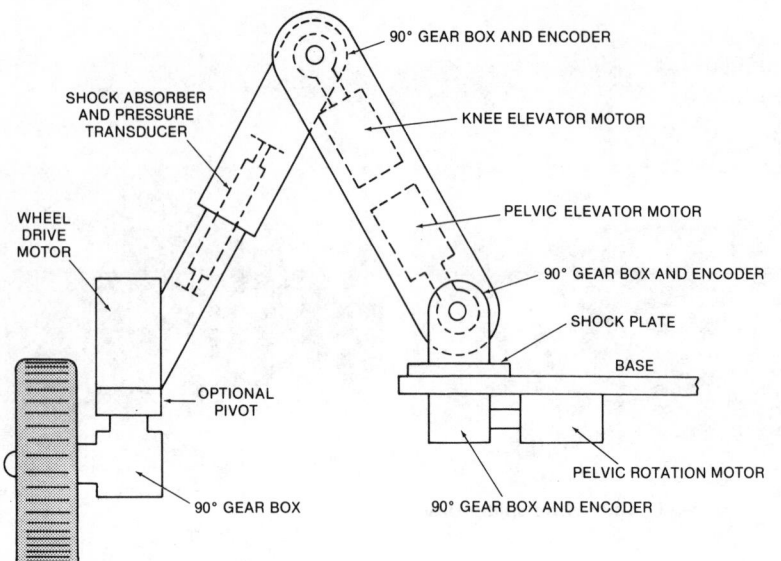

Fig. 3-32. A simplified "spider" leg assembly.

Courtesy Ohio State University.

Fig. 3-33. The OSU Hexapod.

Hexapod is kept tethered and is made to walk short distances over obstacles. One of the forelegs is instrumented with a set of strain gauges and is used for the more complex functions. This is in keeping with the very well proven research axiom of simplifying the problem to the basics.

As with most basic research, the most promising thing about the Hexapod is not the device itself, but the new concepts that are being generated and/or verified during its development. One of the most fundamental concepts that has been studied is "active compliance." This concept arises because position control alone is not sufficient for a walking machine since the sensor systems of the robot will not likely know the exact level of the surface of the ground at the point of contact of each foot. Additionally, the surface firmness may vary greatly under different feet. For these reasons, the control algorithm must include the force being exerted on the surface as part of the feedback loop. This is done by adding a second term to the error feedback signal of the control loop.[5] For a rotational joint, the simple position error is

$$ERROR = K \times (\theta_c - \theta)$$

where,
θ_c is the commanded angle,
θ is the actual position,
K is the feedback gain.

This can also be stated as

$$ERROR = K \times E_a$$

where,
E_a is the angular error.

With active compliance, a torque term is added and the equation becomes:

$$ERROR = (K_a \times E_a) + (K_t \times E_t)$$

In this case, the angle and torque are both commanded, and the total error is taken to be a combination of the torque error and the angular error. To understand this consider the case of the vertical plane hip joint of a leg as the leg approaches contact with the ground. If the angular error reaches zero and the foot has not yet touched the surface (the ground is lower than the robot expected), the leg will not stop moving. This is because there is still an error signal to drive it. This signal is supplied by the torque term. Since the foot has not touched the surface, this term has an error contribution to the whole error. Conversely, if the leg touched down prematurely, it would not move all the way to the commanded position as the load torque term would go positive after the torque target was passed. This means that the robot trades off torque for position. This same process is used in the velocity and acceleration control loops of the robot. The ratio of the gains of the two terms (K_a and K_t) gives the compliance ratio. McGhee has found that this factor should best be adjusted for the roughness of the ground. It should be noted that compliance is necessary in the horizontal plane of control as well as in the vertical plane.

Other interesting facts have come to light during the development of the Hexapod. According to McGhee, a walking robot should be *more* efficient on soft surfaces (such as sand and mud) than a rolling machine. This is because rolling machines (treads included) generate a *bow-wave* effect. This continuous displacement of material all along the path of motion represents a significant power loss. In actual tests, however, walking machines are far less efficient. This is due to several causes, including the use of worm gears in the joints. McGhee has noted that for a walking robot to be efficient, it must recover the kinetic energy from a limb as it slows the limb's motion relative to the body (especially in the unloaded arc of its movement). Worm gears and most common hydraulic controls are not capable of doing this. Additionally, neither of these is very efficient in the first place! Direct-drive motors with back emf braking and power recovery may provide a partial answer to this problem in the future.

It should be noted that the translation between the desired cartesian forces, motions, and positions and their angular counterparts is nontrivial. The term *proprioceptive* is used to describe the joint control, and the term *exteroceptive* is used to describe the vector ground reaction forces. McGhee used Jacobian transforms to develop the relationships between these two systems, but the explanation of these is beyond the scope of this discussion.[16]

The evaluation of the Hexapod suitability is the same as that of the walking mode of the Hybrid Spider Drive. For this reason and to prevent distraction from the more important previous discussion, it will not be repeated here.

The CMU Planar Hopper

At first glance, the two dimensional pogo stick being studied at Carnegie-Mellon[17] might seem absurd. This is distinctly not the case. The study being done by Marc Raibert and his associates involves the most fundamental but as yet not well understood dynamics of motion. The similarities of the legged walkers discussed thus far to insect/spider locomotion goes well beyond external appearance. Most insects walk in a statically stable manner, which is to say that if you froze the position of an insect in mid stride, it would remain standing (assuming it did not fall over from inertia). This is also true of the multilegged walkers. Higher vertebrates move in a much different and more efficient manner. If one froze the position of a walking man in mid stride, he would be totally unstable. The same thing is true of even a walking dog or horse, to say nothing of a running animal.

The study being done on the one-legged hopper is fundamental to understanding the mathematics of such motion. To simplify the problem, the original Planer Hopper shown in Fig. 3-34 was designed to hop in two dimensions only. This was accomplished by placing the device on a sloped "air" table. Air-cushion bearings provided a nearly frictionless support while the configuration constrained the robot leg to a single plane. More advanced versions, including a three-dimensional hopper are now being developed.

Raibert notes that biological locomotion systems store energy in three different ways: (1) in the kinetic energy of motion, (2) in the potential energy of vertical position, and (3) in the elastic energy of muscles and tissues.

Each articulated portion of the leg has its own combination of forms of energy. One of the unique features of the Hopper is the inclusion of spring elements into the leg. This gives the leg the capability of the third form of energy storage mentioned above. This also brings resonance effects that have not been significant in the rigid-leg designs to date. When these effects are better understood, it is hoped that the knowledge can be applied to multilegged robots. Since the "Hopper" is not intended as a load carrying carriage system, it would not be appropriate to subject it to the suitability evaluation.

The Bizarre and the Sublime

In closing this topic, it is interesting to consider a few of the more bizarre schemes that have been mentioned. Moravec has suggested

Courtesy Carnegie-Mellon University.

Fig. 3-34. The CMU planar hopper.

the futuristic concept of "Bush Robots."[18] These robots would resemble a plant in that they would have one main trunk that would have a joint at one or both ends with two smaller limbs attached to each joint. At the end of each of these limbs would be a joint with two even smaller limbs, etc. The smallest of these limbs would be endowed with a high degree of touch and perhaps heat sensitivity. Such a robot would be able to configure its shape to almost any volume. Moravec sees this as particularly useful for performing repairs in tight places.

Another interesting extreme might be called "Amoebic Propulsion." This system would consist of numerous elastic bags glued to each other in a cluster. The outer surface of each bag would be coated with a pressure-sensitive (touch) grid. The electronic controls, some valves, and a small hydraulic pump would be mounted in the center of this blob. Motion would be accomplished by transferring fluid from one bag to another. While this system presents some interesting problems in navigation and load carrying, it should be pointed out that it has excellent traction and an extremely low center of gravity!

These amusing concepts have been mentioned for a very important reason: we should not be embarrassed to think about and discuss the most unconventional and unlikely of possibilities. When such ideas are the topic of conversation, unexpected and truly useful inspirations sometimes result. These ideas may be only remotely connected to the concept that "seeds" them. Robotics is an enormous field and its possibilities have only begun to be explored. If we could see the realities of the next decade, they might seem even more unlikely than the ideas I have just mentioned!

REFERENCES

1. Long, J.E. and Healy, T.J. "Advanced Automation for Space Missions." University of Santa Clara, Santa Clara CA.
2. Balmer, C. "Avatar: A Homebuilt Robot." *Robotics Age*, Peterborough, NH January/February 1982.
3. Hoffstatter, G. "Ambulatron." *Robotics Age*, Peterborough, NH January/February 1982.
4. Beiser, A. *Applied Physics.* McGraw-Hill Book Co., NY, NY.
5. McGhee, R.B., Olson, K.W., and Briggs, R.L. "Electronic Coordination of Joint Motions for a Terrain Adaptive Robot." Society of Automotive Engineers, Inc., Warrendale, PA.
6. Klein, C.A. and Briggs, R.L. "Use of Active Compliance in the Control of Legged Vehicles." IEEE Transactions on Systems, Man, and Cybernetics, Vol. SMC-10, No. 7, July 1980.
7. McGhee, R.B. and Iswandhi, G.I. "Adaptive Locomotion of a Multilegged Robot over Rough Terrain." IEEE Transactions on Systems, Man, and Cybernetics, Vol. SMC-9, No. 4, April 1979.
8. McGhee, R.B. "Future Prospects for Sensor Based Robots" from "Computer Vision and Sensor Based Robots." Plenum Publishing Corp., 1979.
9. McGhee, R.B. *Personal Communication*, February 1982.
10. Paine, G. "The Automation of Romote Vehicle Controls." Proceedings of 1977 Joint Automation Control Conference, San Francisco, CA.

11. Moravec, H.P. "The CMU Rover." Carnegie-Mellon University, June 22, 1981.

12. Von Daniken, E. *Chariots of the Gods.* Bantam Books, New York, NY, 1970.

13. La, W.H.T., Koogle, T.A., Jaffe, D.L., and Leifer, L.J. "Microcomputer Controlled Omnidirectional Mechanism for Wheelchairs." *1EEE*, "Frontiers of Engineering in Health Care" CH1621-2/81/0000-0326, 1981.

14. Weinstein, M.B. *Android Design.* Hayden Book Co., Rochelle Park, NJ, 1981.

15. Gurfinkle, V.C., et al. "Model Six-Legged Walking Machine with Supervisory Control." Report No. 2036, Institute of Mechanic, Moscow, USSR (in Russian).

16. Whitney, D.E. "Resolved Motion Rate Control of Manipulators and Human Prostheses." Vol. MMS-10, *IEEE Transactions on Man Machine Systems,* June 1969.

17. Raibert, M.H., et al. "Dynamically Stable Legged Locomotion." The Robotics Institute, Carnegie-Mellon University, Pittsburgh, PA, September 1981.

18. Moravec, H.P. "Bush Robots." The Robotics Institute, Carnegie-Mellon University, Pittsburgh, PA, September 1981.

MANUFACTURER REFERENCES

M1. Nuclear Research Center
Karisruche, Germany

M2. International Robotics Inc.
316 E. 53 Street
New York, NY 10022

M3. R B Robotics Corporation
14618 West Sixth Avenue
Suite 201
Golden, CO 80401
(303) 279-5525

Chapter 4

Passive
Vision Systems

Providing the robot with an accurate and useful vision system is undoubtedly one of the most challenging design problems associated with robotics. Like all of the other technologies required in robotics, vision systems have improved rapidly in the past decade. Charge-coupled-device (CCD) cameras are being used increasingly in robotic applications, and the processing of two-dimensional high-contrast images has become almost commonplace in industrial robots.

Unfortunately, the problem of processing an image in three dimensions has only recently been addressed. The ability to process three dimensions is extremely important in systems where the distance between the camera and the object of interest is variable or unknown. Robotic arms with cameras mounted near the manipulator end and mobile (or partially mobile) robots are good examples of systems requiring this capability. In such systems, information from the vision system must be correlated to the information from the other sensory systems of the robot, correlated to the information processed in the recent past, correlated to the information from the navigation system (if mobile), and correlated to the tasks assigned to the robot. As formidable as this correlation might seem, it has already been accomplished to some extent in at least two robot research projects (Stanford Research Institute's "Shakey" and Moravec's "Rover").[1] Although both of these robots fell short of being practical, they have laid the groundwork for the systems that will follow. As important as any aspect of this advanced research, is the fact that it

171

points out the type of optical and computer hardware that will be required to perfect these systems.

THE VIDICON CAMERA

There are, at present, two types of cameras that are finding significant application in robotic vision systems. The first of these types is the common vidicon.[M4] The principle used by this device has remained unchanged since the invention of practical television. As shown in Fig. 4-1, the device is essentially a vacuum tube with a heated cathode, control grids, and a special optical target.[2] The target is effectively an array of microscopic capacitors with one common plate. Each capacitor loses charge at a rate determined by the light incident on it (and other factors). The target common plate is charged to a potential slightly (approximately 20 volts) more positive than the cathode. An electron beam is emitted from the cathode and is accelerated by a modest (several hundred volt) positive charge on the surrounding grid #3. The device uses magnetic focusing and deflection to scan the beam across the image target. As the beam strikes one of the *photo-capacitors*, it will recharge it. The recharge current will be proportional to the amount of charge lost since the previous scan and thus to the incident light. The current signal from the common lead of the capacitors is collected by a contact ring and is available as the video output.

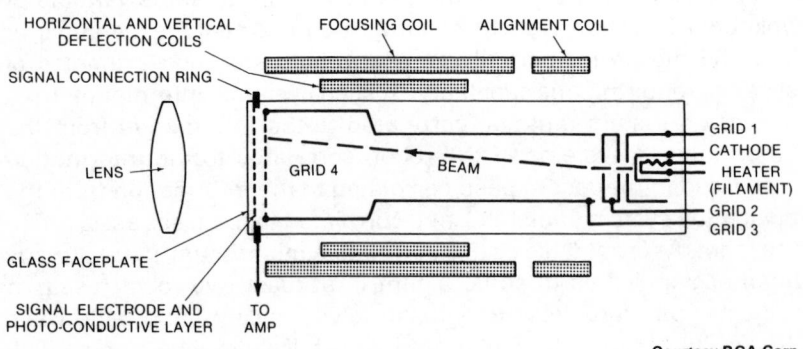

Courtesy RCA Corp.

Fig. 4-1. Construction of a typical vidicon tube.

PLUMBICONS AND OTHER RELATIVES OF THE VIDICON

There are a number of other imaging tubes besides the vidicon that find application in robotic and automation systems. Each of these types has one or more advantages in the particular field of application. The Plumbicon,[M3] for example, does not have the image retention of the vidicon. This tube is therefore able to operate with a fast-moving image without blurring and thus is often chosen when this characteristic is critical. Other systems require extremely high resolution or optical sensitivity. The Ultracon and Newvicon,[M4] for example, provide a sensitivity improvement of almost a factor of ten over the vidicon. As might be expected, the spectral response of these tubes varies rather significantly, and this will be discussed in a later section of this chapter.

Image-tube cameras are also available in a wide variety of housings. The examples shown in Figs. 4-2 and 4-3 are only a few of the available housing configurations. The wide variety of application-oriented housings is a result of the relatively long historical use of these cameras. A solid-state (CCD) camera is also shown in Fig. 4-4 for comparison purposes.

COMPOSITE VIDEO

The control electronics found in most small vidicon cameras generate what is known as composite video (see Fig. 4-5). This is done by sweeping the beam from left to right across the image target 15,750 times per second. The first sweep of a field takes place at the top of the image, and every successive sweep starts slightly below the previous one. At the end of a sweep, a short horizontal synchronization pulse is generated to signal the receiver that a new line is beginning. If this pulse and the vertical synchronization pulse are combined with the video signal, the resulting signal is called *composite video*. Each new field is separated from the previous field by a vertical synchronization signal. There are 240 scan lines in each field, not counting the lines (22 ½) that occur during the vertical synchronization signal. This type of scanning is referred to as *raster scanning*.

The vertical-synchronization signal may be serrated by either normal- or double-rate horizontal-sync pulses. If double-rate pulses are used (RS 170), they are odd in number causing a shift of one-half

(A) A vidicon camera (500-line resolution).

(B) A camera with auto iris (left) available with vidicon, newvicon, or ultracon.

Courtesy RCA Corp.

Fig. 4-2. Typical image tube cameras.

Fig. 4-3. An image tube camera rated for space and military applications. This unit is pressurized with dry nitrogen.

Fig. 4-4. A CCD camera.

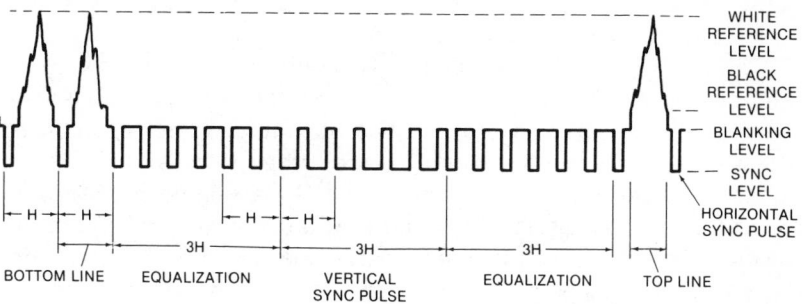

Fig. 4-5. Composite video with interlace (field 1).

line on every other field. This technique causes the lines of every other field to be interlaced with those of the previous field and is thus called *interlacing*. When such a scheme is used, two fields are required before the full resolution of the picture has been transmitted, and thus two fields are said to make up a *frame*. Many robotic systems do not require the added resolution provided by this technique, and lines are often treated as if they were not interlaced. It is, of course, possible for the video processor to use the additional resolution achieved by interlacing, but it is fairly common not to. Unfortunately, if all fields are treated as the same image, the edge of an object may appear to vibrate vertically at 60 Hz. If this represents a problem, the controller can either ignore odd fields, or (in some cases) the camera can be set to a noninterlace mode. In any given system, the trade-off will depend on the requirement for resolution versus the processing speed. An image with twice the resolution will normally require twice the processing time.

Unfortunately, image tubes have at least four serious disadvantages in robotic applications: distortion, power consumption, the requirement for high voltages, and size. In addition to lens distortion (a problem in all cameras), image tubes suffer from distortion related to their magnetic scanning technique. Since the vertical- and horizontal-deflection circuits are independent, they are never perfectly matched nor are they completely uniform across the tube. They are also subject to stray magnetic fields from ferrous metals and electrical currents. If there is a significant error between the vertical- and horizontal-deflection factors, an object will appear to change shape as it is rotated with respect to an axis pointed at the camera lens. This distortion can easily destroy the effectiveness of the techniques that will be discussed later in this chapter. The distortion problem may be severe or minor, depending on the quality of the camera and the environment in which it is used. As an example, an image-tube camera might be very difficult to use near an arc welder since the arc currents would generate very large magnetic fields.

In many present-day systems, vidicons (and their relatives) are still being used because they are less expensive than CCDs with comparable resolution. Other practical advantages that have allowed these devices to continue to hold a place in the robot market are that a lot of interface equipment is commercially available for use with composite video signals, and that the output of a vidicon camera can be monitored directly by an inexpensive video monitor. These are not inherent advantages of the vidicon itself but of the state-of-the-art, available hardware that incorporates it.

CHARGE COUPLED DEVICES (CCD'S)

Because of the disadvantages associated with the use of vidicons mentioned above, charge coupled devices are rapidly displacing them in many areas of robotics. Until recently, the lack of high-resolution CCDs made them inadequate for many applications. The resolution of a CCD is rated in pixels. A pixel is a single photodetector element. CCDs are only now becoming commonly available with resolutions of 256×256 pixels and better. Previously, the best resolution available was 100×100 pixels. This format makes CCDs awkward to interface to computers (since it is not a power of 2), and it is even more awkward to interface to standard raster scan monitors (which use 240 scan lines per field). Compromises with this problem will be discussed later in this section.

To understand the difficulty that a robot might experience in trying to identify a particular object in an image, try to identify the object shown in the various resolution images of Fig. 4-6. It is not difficult to understand that a camera with a resolution of less than 64×64 pixels would have limited application in even mildly demanding visual tasks.

Typically, a CCD sensor is a large array of photodetectors (Fig. 4-7), coupled to external circuitry by an array of *FET (Field-Effect Transistor)* switches, and one or more *analog shift registers* called a *BBD (Bucket-Brigade Device)*. The simplified diagram shown in Fig. 4-7 is that of a 100×100 pixel CCD used in the Reticon MC 520 camera.[M1]

To minimize interconnections to the CCD component, an internal counter sequentially connects successive rows of detectors to one of two bucket-brigade devices (also called transports). When the charge of a row of detectors has been loaded into the corresponding stages of a BBD (odd-numbered rows are transferred to one BBD and even-numbered rows are transferred to the other BBD), the charge can be shifted out by simply clocking the BBD. The use of two BBDs allows one BBD to be loaded while the other is being shifted out. With the addition of some simple camera electronics, control can be accomplished with three signals. Two of these signals are generated by the computer control electronics, and one is supplied from the CCD camera electronics to the control electronics. One of the signals from the control electronics (MIC) advances the row counter and synchronizes the transfer of charges to the BBD stages, and the other signal to the camera (MCLK) shifts the BBD video signal out to the processor. The third signal (EOF) indicates that an end-of-frame condition exists

177

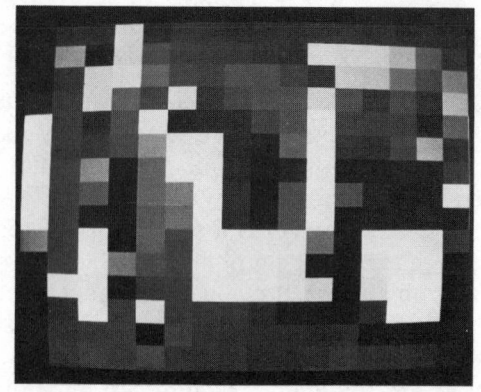

(A) 16 × 16.

(B) 32 × 32.

(C) 64 × 64.

Fig. 4-6. The effect of the

(D) 128 × 128.

(E) 256 × 256.

(F) 512 × 512.

Courtesy Colorado Video Inc.

number of pixels on resolution.

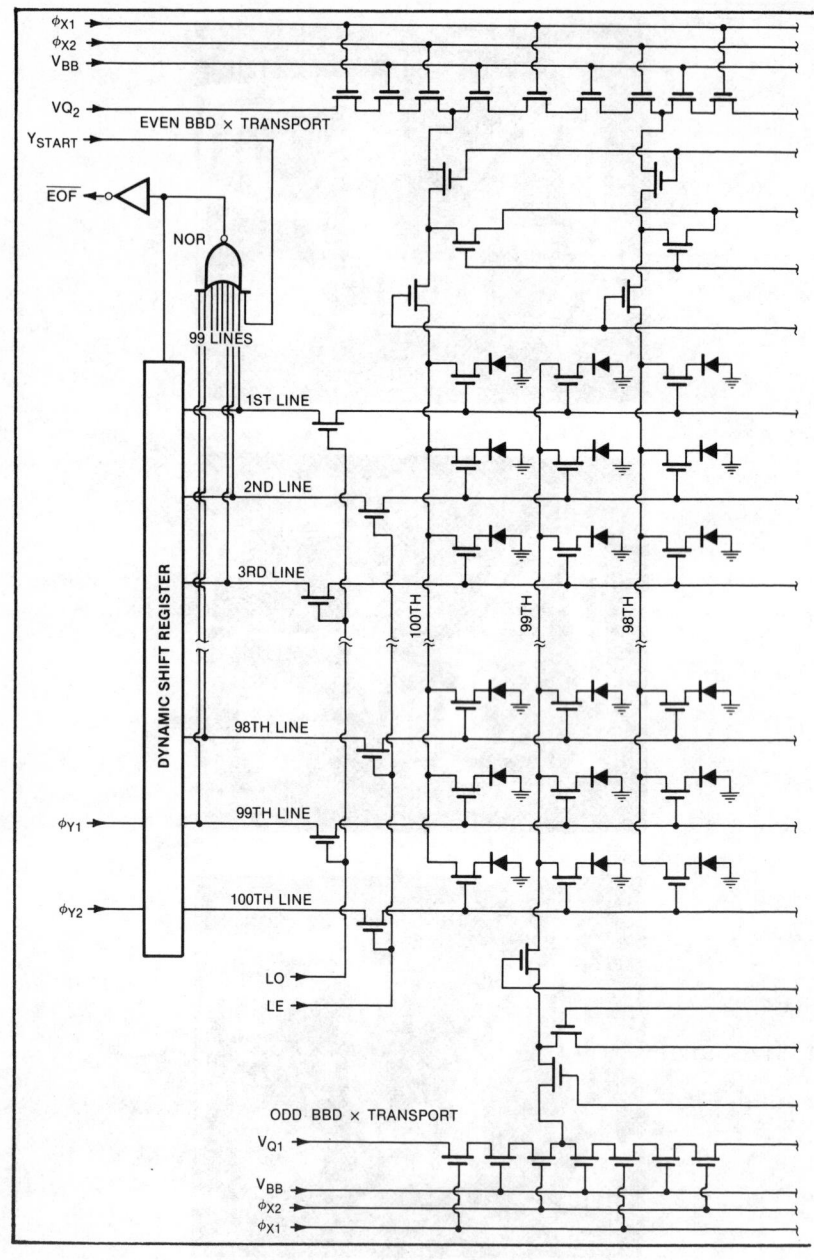

Fig. 4-7. CCD

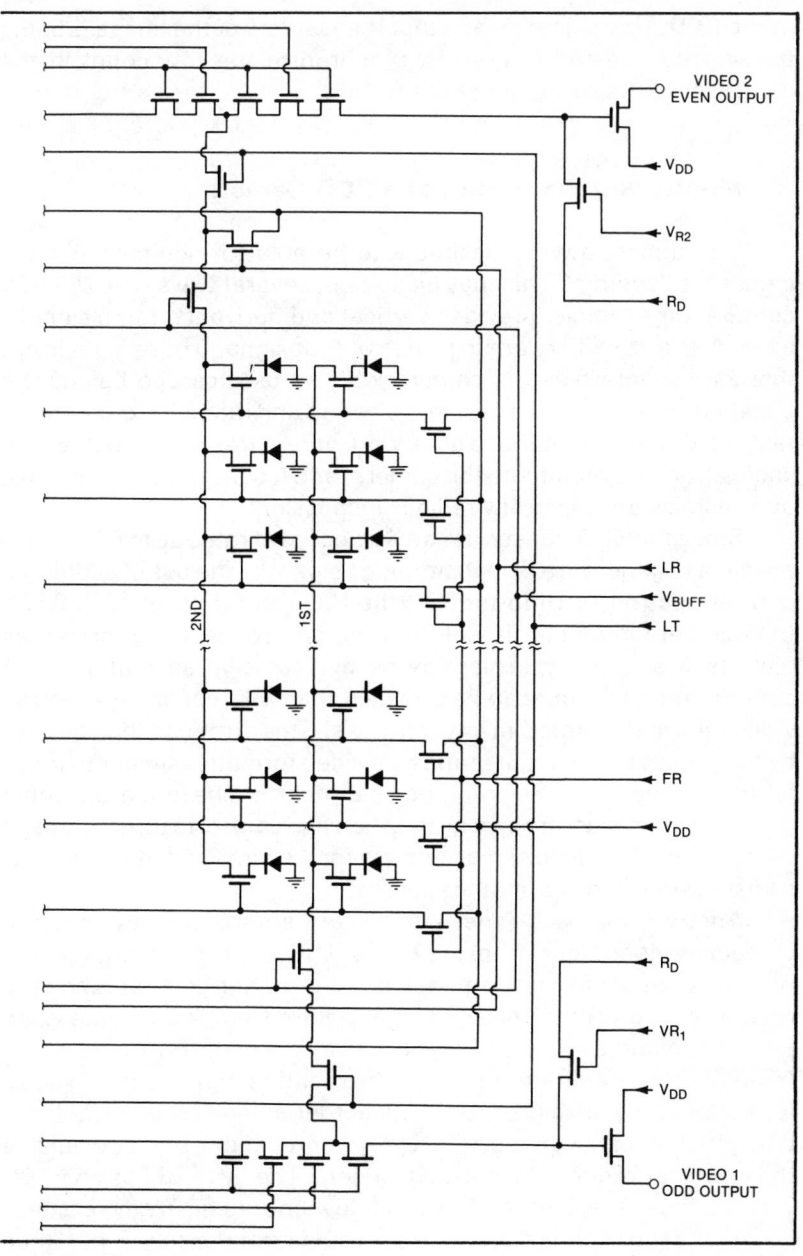

VIDEO 2
EVEN OUTPUT

V_{DD}

V_{R2}

R_D

LR

V_{BUFF}

LT

FR

V_{DD}

R_D

VR_1

V_{DD}

VIDEO 1
ODD OUTPUT

2ND

1ST

Courtesy E.G. & G. Reticon.

architecture.

181

in the CCD. This signal means that the last row of the image is being transferred out and is used to synchronize the row count in the controller to the row counter of the CCD.

Monitoring the Output of a CCD Camera

It is almost always desirable to be able to monitor what the camera is "seeing." This may be done in several ways. The MC 520 camera, for example, provides vertical and horizontal ramp signals, as well as an auxiliary analog video output signal. These signals are intended to interface the camera with an oscilloscope having the capability of external horizontal deflection and intensity (Z axis) modulation. While this interface does work, it requires three BNC coaxial cables from the monitor to the camera, and it does not allow for ease of transmission (especially radio transmission).

Since the BBD transports can be clocked at rates up to 5 MHz, it is possible to generate a composite video signal by the use of additional control electronics. Unfortunately the 100-row format of the MC 520 CCD camera does not fit well into the composite video format (as discussed previously). This may be overcome in several ways. A camera interface unit can be provided that will repeatedly read the video information into a memory array(s). This array can then be read by the processing computer and by a video formatter similar to those found in computer video graphic display boards. The formatter would generate the necessary synchronization and control to turn the image into a composite video signal suitable for viewing on a monitor or for transmission by radio or other means.

Another approach (one used by the author) for developing a composite video signal from the MC 520 CCD camera (or any camera with an unusual format) is to have the formatter generate the synchronization and blanking signals as required and "pump" the video out of the camera so as to fill in the active area of the horizontal lines. In addition to generating the synchronization signals, this type of formatter would also produce blank border areas at the top, bottom, left, and right of the image area. In the center of this border would be the video as produced by the CCD camera. The MC 520 has only 100 vertical lines of resolution. This is too few lines to fill in any reasonable area of the screen, but 200 lines would format nicely. Since lines cannot be read out twice in a row, the formatter will either have to buffer a line and alternately read it back out, or it will have to intersperse blank lines with active lines from the camera. The second of these approaches works fairly well and is relatively inexpensive. As the data is "pumped" from the camera, it must also be placed into memory for access by the computer.

Disadvantages of CCD Cameras

In keeping with Murphy's law, CCD cameras have several disadvantages. Probably the most significant problem of the CCD is that the optical sensitivity cannot be controlled as easily as can that of the vidicon. The sensitivity is directly proportional to the time that the image is allowed to strike the detectors between transfer operations. When excessive charge has accumulated, the detector will spillover causing the image to bloom and distort. In CCDs, the most common result is for the excessive charge to "punch through" the array switches or the elements of the transport register causing severe vertical distortion of the image. This effect is sometimes referred to as *barrier modulation.* As a result of barrier modulation, a CCD camera viewing a bright point of light might produce an image that is a vertical white bar with a slight bulge at the point of light.

To prevent overexposure (blooming) in the presence of bright optical images, either the device must be clocked faster, or the mechanical aperture must be closed (stopped) down. Clocking the device faster is not usually desirable because the image processor will normally be working as fast as possible to provide the other systems with the image information. Since this information is transferred at a preset rate, if the scan clocking rate is increased, the image processor would have to occasionally discard an image. The result of discarding an occasional image (frame) would be an apparent lurch in the motion of objects tracked by the computer. It is possible to double the clocking rate and throw away every other frame without causing velocity distortion (lurching), but this procedure does not offer very fine control.

Stopping down the mechanical aperture of a CCD camera is presently the most common method of sensitivity control. This requires a mechanical actuator that is inherently more likely to fail than the camera itself. Future developments in CCD cameras will likely include the incorporation of an electronic sensitivity control. Whether this control will consist of an addition to the CCD device itself, or whether it will consist of a separate electro-optical element (such as a liquid crystal device or a Kerr cell) remains to be seen.

In applications where the camera will be exposed to very bright images, one or more neutral-density filters may be attached to the end of the lens assembly. Some systems even offer motorized systems for inserting and removing the filters, but this can disrupt the robot's vision processing and is thus a solution of last resort.

Spectral Sensitivity of the CCD Versus the Vidicon

As shown in Fig. 4-8, the spectral sensitivity of the CCD is much different than that of the vidicon. The CCD has greatly increased sensitivity in the near infrared region and a flatter curve, but falls below the vidicon at wavelengths shorter than blue light. The robot designer with security applications in mind will find this very interesting. The spectral response of the CCD is such that it can "see" the blackbody radiation from objects warmer than approximately 600 kelvins (327° Celsius). There is thus the possibility of using this sensitivity to detect hot objects that would otherwise appear quite normal to the human eye. Additionally, a security robot using CCD vision might operate in apparent darkness by using an ir (infrared) light source. Solid-state light-emitting diodes (LEDs) and lasers are commonly available with emissions in the 900-nm wavelength range.

It should be noted that the curves of Fig. 4-8 are *normalized.* The actual peak sensitivity of the Ultracon, for example, is many times

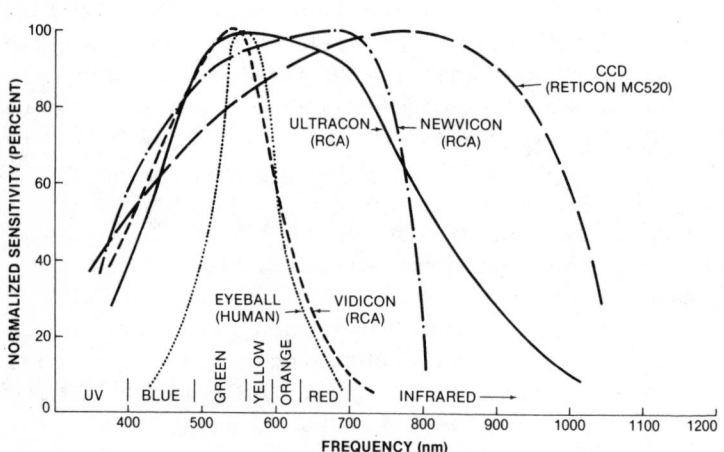

Fig. 4-8. Normalized spectral response of various imaging devices.

greater than that of the vidicon. It is also interesting to note the rather poor spectral response of the human eye.

OTHER SOLID-STATE CAMERA TECHNOLOGY

Other types of solid-state cameras are available, such as those based on the Reticon R0720B circular detectors. These devices can be expected to continue to improve in resolution, price, variety, and performance. For those interested in inexpensive experimentation, a solid-state camera can be constructed from a dynamic RAM chip. (This is described in Martin Weinstein's book *Android Design*.)[4] Periphicon offers relatively inexpensive cameras based on specially modified dynamic RAM chips.[M5] Since all solid-state cameras are essentially arrays of photodetectors, the problems encountered will usually be similar to those discussed under the previous section on CCDs.

OPTICAL CONSIDERATIONS

No matter which camera is chosen, a lens assembly must be selected to provide the characteristics required for the application. One of the primary considerations is the viewing angle of the camera. This angle is determined by the *focal length* of the lens. While most lens assemblies are composed of several optical elements, their focal length is still specified as if they were simple convex lenses. The focal length is the distance from the equivalent center of the lens to the plane of focus. The plane of focus is the plane where an image at a very long distance would focus. The relationship between focal length, optical target (sometimes referred to as format) dimensions, scene distance, and scene coverage is given in Fig. 4-9. The standard target sizes for most image tubes are (0.500 × 0.387) and (0.347 × 0.260) inches.[11] If the viewing area of the robot is too small, the robot may not be able to see enough of an object to identify it. Conversely, if the area is too large, the loss of resolution may cause poor results.

Another factor of importance is the *f-number* of the lens. This number is sometimes referred to as the speed of the lens, since it relates to the shutter-time requirement in photographic cameras. The value of the f-number is related to the ability of the lens to collect light, and is determined by

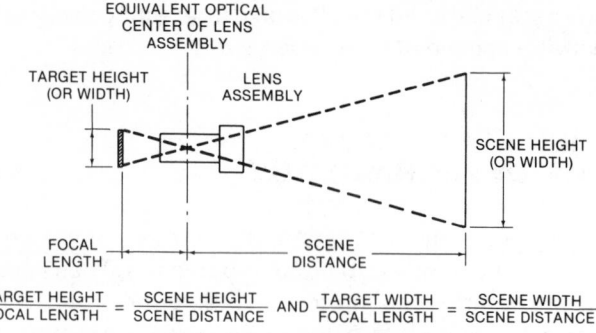

Fig. 4-9. The relationship of focal length to field of view.

$$\text{f-number} = \frac{\text{focal length}}{\text{lens diameter}}$$

A lens with smaller f-number will collect more light, and will be capable of a narrower *depth of field*. More will be said about the importance of this in a later section of this chapter.

DIGITIZERS AND PREPROCESSORS

To the casual observer, it might seem that the approach to digitizing an image is obvious; simply digitize all pixels to the data size (8 bits, 16 bits, etc.) of the processor being used and store the values as a giant array. This is in fact done on many systems. Unfortunately, the amount of data represented by such an image is extremely large, and the speed at which it is changing makes for an extremely large processing task. For example, an array of 256 × 256 pixels, where each pixel is represented by a single 8-bit byte, would require 65,536 bytes of memory for a single image. If this image were to be buffered in such a way as to allow the processor to analyze one image while another was being collected, the requirement would be twice as great. Since 65,536 bytes represents the entire address space of many 8-bit microprocessors, the problem is obvious. Memory size is not the worst problem either. Data from such a video source would be arriving at a rate of almost 4 million pixels per second (assuming 60 frames per second). Only the most expensive and high performance computer can process video data at this resolution and speed. Common microprocessors such as the 6502, Z80, and 8085 are not nearly fast enough for this data rate. The addition of auxiliary mathematical processors (called *array processors*) to modest minicomputers or advanced microprocessors (such as the MC68000) can sometimes provide the required computational power, but the resulting system is still expensive.

Because of the speed limitations of computers (especially microprocessors), it is often desirable to include dedicated hardware in the digitizer to preprocess the video information. Since the operations to be done on each pixel are usually identical (or are of limited variety), this logic does not need to be program driven. Even though it is not program driven, preprocessing logic is usually connected to the processor by interface ports that allow the processor to control certain parameters and operational modes. There are many possibilities for configuring such preprocessing logic, and a few methods will be mentioned in the following pages. The choice of the preprocessing logic configuration is crucial to the efficient operation of a robot vision system and must be based on the tasks that are to be expected of the robot.

Frame, Row, Column, Window, and Pixel Grabbers

Obviously, both processor speed and memory requirements can be reduced if the digitizer does not place the entire image into memory at any one time. For this reason, digitizers are often designed to convert only part of the image field. These types of digitizers have been given descriptive names accordingly.

A system that digitizes an entire image into memory at one time is called a frame grabber. At the opposite extreme, a digitizer can consist of nothing more than a pixel trap that waits for one pixel to be read. The horizontal and vertical location of this pixel is supplied by the processor by writing it into two output latches connected to the trap logic. When the specified pixel is read, the result (either a gray-scale value or a binary value) is saved in a latch that can be read by the processor. The processor is signaled by either an interrupt or a handshake signal that the requested pixel is available. Unfortunately, if random access to pixels is desired, only one pixel can be read during each frame (1/30 or 1/60 of a second). This destroys the usefulness of the simple pixel grabber for any but the slowest applications. Colorado Video Inc. provides modular image acquisition systems in which one module is the pixel grabber.[M6] These systems can be configured with the pixel grabber operated directly by the process (as described above), or with the pixel grabber connected to a frame buffer or a direct memory access (DMA) controller. Such systems offer "building-block" flexibility combined with high performance.

Between the fast frame grabber and the slow pixel grabber lie the natural compromises of line and column grabbers, and window grabbers.[M7] These digitizers do exactly as their names indicate, collecting only the values of pixels in one specified column, along one line, or in an area of the screen that has been defined by a set of upper/lower, and left/right limits. Some digitizers can operate in more than one of these modes, as commanded by the processor. For a robot with a relatively simple task, such as following a seam in two pieces of metal, such digitizers are often satisfactory. They may be of limited use, however, in more demanding applications.

The Random Multiwindow Digitizer

An alternate approach to a single, full-resolution frame is a system having a large number of windows (all within the frame), each of which is programmable in height and width. Under this approach, windows may overlap. This digitizer provides the processor with either the total or average value of the pixels falling into each window. The random-window digitizer is fast, but not very flexible from a programming standpoint. This limitation is because it cannot easily deal with problems such as variable-image rotation and range. For this reason, this digitizer is better suited for conveyor-belt-type systems (where range and orientation are fixed) than it is for most robotic applications.

Variable-Resolution Windowing

Several other alternatives based on the windowing concept can be implemented. For example, a digitizer can be made programmable in such a way that it can change the effective window size at the same time it changes resolution, thus minimizing and simplifying the data gathered. The memory image produced by such a digitizer would be constant (e.g., 32×32), but would represent an area of varying size and resolution. Such a digitizer could be made to return the full resolution image of a 32×32 pixel window, for example, or it could be made to average the pixels falling into each of 1024 (32×32) subwindows that together form a fuzzy image of the entire frame.

Starting with the lowest resolution, such a digitizer could be used to quickly zoom in on an area of interest by changing to consecutively smaller, higher-resolution windows. If, on closer inspection, an area was found to be the incorrect one, the digitizer could move back out and look for another candidate. Additionally, this type of digitizer could allow the robot to focus on a small image (such as a man walking in the distance), while alternately watching the whole scene (in low resolution) for movement. This type of operation bears a significant resemblance to the peripheral vision of humans. The use of this format could also reduce the bandwidth requirements for television transmission to a base computer or remote-control console.

Unlike line and column grabbers, the variable-resolution method is one of the few compromises that may find serious use in demanding robotic applications. In his program for locating known reference points in consecutive images, Moravec used a window resolution scheme very similar to that described in the preceding paragraph.[1] The method improved the reliability of the correlator program while reducing false identifications. Moravec used a full-frame image digitizer, and the resolution changes were done in software. The process did not, therefore, offer the speed advantage that a variable-resolution digitizer could. This was not important to his system since other functions accounted for the major part of the processing time. Moravec's approach did however serve to point out the possible usefulness of such a digitizer.

A variable-resolution digitizer would not be appropriate for every vision requirement, especially where a processor must analyze the entire area of a relatively large object at full resolution and in a single operation. It does, however, offer a reasonable solution to problems that require low-cost hardware and fast processing such as mobile robotic devices.

FULL-FRAME IMAGE PROCESSING

Since many industrial robotic applications require the performance of a full-frame digitizer, the methods of processing these images will be the only ones considered in depth in the following pages. This is also dictated by the fact that most serious research in video processing to date has been confined to the frame-grabber-type digitizer.

Before the formal processing of the image begins, it is advantageous to simplify it as much as possible. This image conditioning usually consists of either converting the image to a black and white (binary) image (Fig. 4-10) or of converting it to a collection of edges. In some cases, binary conversion is an intermediate step to edge determination.

Binary Conversion

Binary conversion could be accomplished by first storing the gray-scale image and then comparing each gray-scale value in the image memory to a set threshold, thus generating a second (binary) image from the resulting data bits. This process would require processor time and is thus not very desirable. This method does, how-

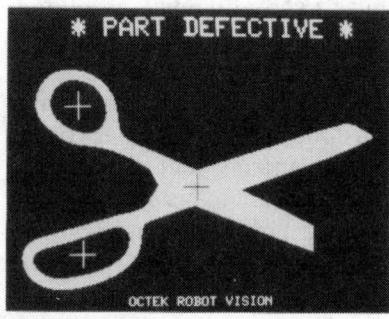

(A) Product inspection.

(B) Process verification.

(C) Process control.

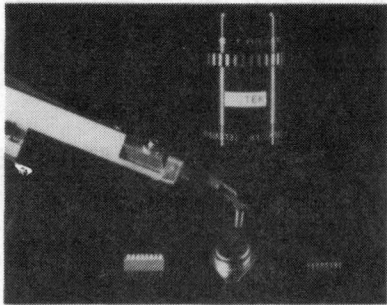

(D) Machine vision.

Courtesy Octek Inc.

Fig. 4-10. Examples of practical vision applications using the "Blob" technique.

ever, allow for scanning the gray-scale values prior to conversion to determine the appropriate value of the threshold.

A more common approach to binary conversion is the one used in the scheme discussed previously. In this case, incoming video data is compared to a threshold voltage by an analog comparator. The resulting bit for each pixel is stored in the image memory. The threshold voltage can be either a fixed reference (from a potentiometer) or a programmable voltage from a d/a converter controlled by the processor. Some digitizers have the ability to change the threshold voltage as a function of the average background intensity in various quadrants of the screen. The resulting black and white image will usually be stored as a number of bits packed into 8-, 16-, or 32-bit words. This packing can be done at the incoming data speed by the camera control interface. The decision of whether or not to pack is determined by the particular computer hardware being used, its address space, addressing modes, instruction set, etc. In some cases, packing will slow processing time; in other cases it will speed it.

Dual-Threshold Preprocessing

A variation of the binary image is accomplished by an input processing scheme called the *dual-threshold* method.[3] The dual-threshold process stores a 1 in a pixel location if the value of the pixel falls between a high and low threshold. If the image pixel is below the lower threshold, or above the higher, it will be stored as a 0. This method serves as a sort of edge generator. The technique is extremely fast and inexpensive. Although this digitizer basically generates only a binary image, it can be made to generate a gray-scale image if the picture is not changing rapidly. This can be done by operating the digitizer in the single-threshold mode and changing the threshold value on successive digitizations of the picture. The more gray levels done in this way, the slower the image can be allowed to change without causing digital "blurring."

The dual-threshold method has the disadvantage that it works well only on objects with rounded edges. Flat objects and objects with sharp edges may not produce any pixels between the two thresholds on one or more of its edges even though one of the two surfaces forming the edge is above the upper threshold, and the other is below the lower threshold. Additionally, the "pseudo-edges" produced by the method may be of various widths, and are not pure pixel *chains*. Like all of the techniques discussed here, selection of the dual-threshold technique must be based on the expected applications of the system. In general, the dual-threshold method is better suited to systems using simpler processing algorithms than will be discussed here.

Edge Detection

A high-contrast (to the background) flat object can be easily converted to a binary image (Fig. 4-10). Once this is accomplished, the edges of the object are known. In many cases, however, edges are not nearly so easy to detect. When this is the case, several algorithms can be used on the gray-scale image to develop an *edge map*. This map would consist of a two-dimensional binary representation of single-pixel-width lines corresponding to the edges of the various objects' silhouettes.

Most of the methods for determining the edge map are based on the derivative (and/or second derivative) of the gray-scale values in two dimensions. This is essentially nothing more than looking at the rate of change of the intensity across the image. Visualizing this process in two dimensions is difficult at first. For this reason, the next few examples will be simplified to one dimension. For those readers not comfortable with the concepts of derivatives, a waveform containing several common features is shown in Fig. 4-11A. These features include a step, a linear ramp, and a curved (raised cosine) slope. For the purpose of this example, analog signals are shown. Notice that because the derivative is the change of the signal in the Y direction (dy) over the change in the X direction (dx) (or dt if the X axis represents time), it is referred to as dy/dx or dv/dt. In the analog representation, all functions are continuous. Thus, if the amount of time over which a significant amplitude change occurs is very small, the derivative is very large. For this reason, the first derivative of the step and the second derivative of the ramp are extremely large and outside of the dynamic range of the circuit. The voltages representing the derivatives would be limited by the power-supply voltages of the circuit.

The peaks of the first derivative of the image correspond to edges as seen by the human eye. The determination of the peak value of the first derivative is easily accomplished by looking for the zero crossings of the derivative of the first derivative or more simply stated the second derivative (Fig. 4-11B). Some algorithms simply look for peak values in the first derivative, but this is mathematically comparable to looking for the zero crossing of the second derivative. Unfortunately,

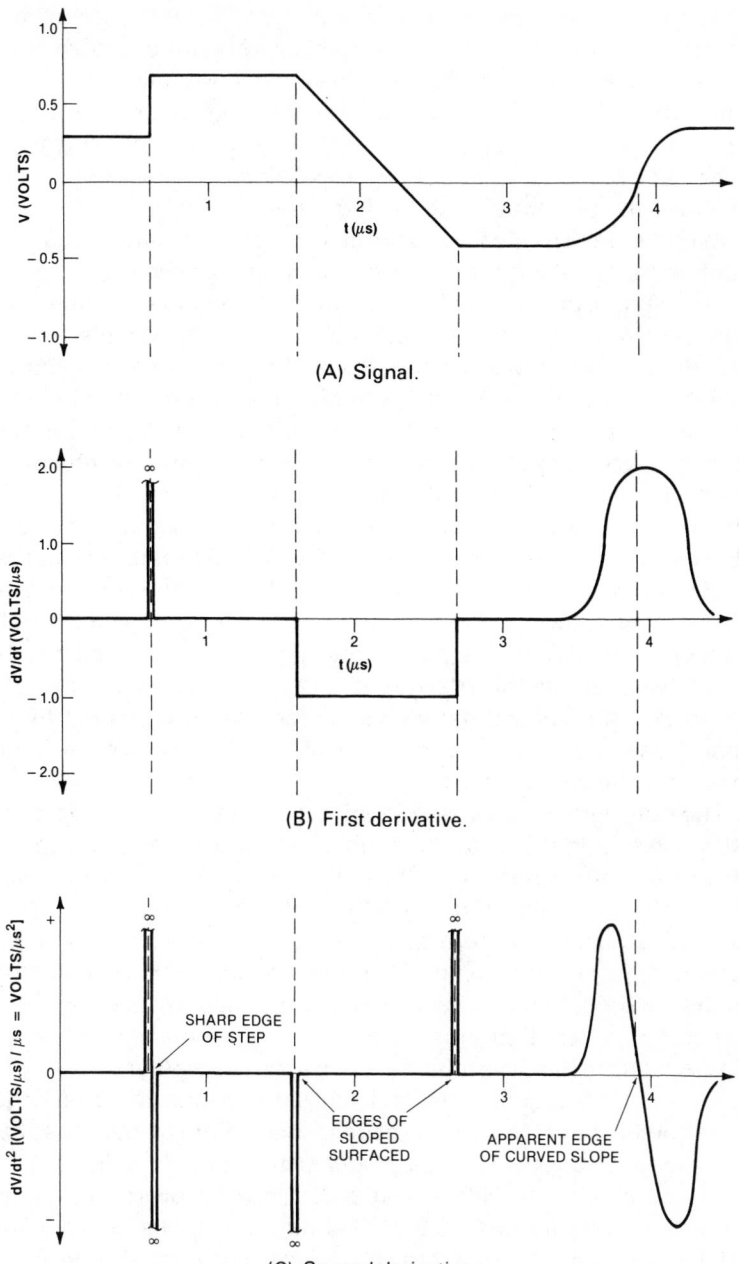

(A) Signal.

(B) First derivative.

(C) Second derivative.

Fig. 4-11. First and second derivatives of common waveform segments.

the process of taking the derivative of a signal is *very* susceptible to noise. Because noise is a random signal (not a continuous function), it introduces many rapid changes in the signal. Fortunately, the effect of differentiated noise is to generate random dots on the image edge map, and since these dots do not tend to form continuous lines they can be reduced or eliminated by processing. As a side note, this problem with noise may make analog transmission of a video signal from a robot undesirable for any purpose other than monitoring (even though other considerations may dictate that it be done).

The next example (Fig. 4-12) is still restricted to one dimension but is somewhat closer to the real problem. In this example, a gray-scale representation of a horizontal scan line of an image is shown. Since the computer's image has a minimum horizontal resolution of one pixel, derivatives cannot become infinite. The apparent step changes at the edge of each pixel are ignored by the processor because it makes only one calculation at the center of each pixel. To help you understand the process of generating derivatives, the equivalent continuous function is shown in Fig. 4-12B to aid the visualization of the process. The apparent edges show up at the zero crossing of the second derivative. Notice that only the sharp edge at the right shows up distinctly as a sharp zero crossing. It may be noticed that the absolute value of the third derivative (not shown) of the signal at the point where the second derivative crosses zeros is related to the distinctiveness of the edge, but use of the third derivative is not necessarily the best solution.

The real world is composed of objects of vastly different sizes and textures. Additionally, a particular object viewed close up will appear to have much smoother edges than it will when viewed at a distance. It is helpful to think of the gradient of intensity across a two-dimensional image in terms of its analogy to temporal (time-referenced) frequency. It can, therefore, be said that an object that appears to be small will have edges with high *spatial-frequency* content. For robots that must deal with diverse images, it is often advantageous to filter the image in such a way as to filter out the finer (higher spatial frequency), more localized edges in favor of larger fuzzier (lower spatial frequency) edges. These more global edges can then be mapped back onto maps of edges found after less filtering.[5] At MIT, researchers E.C. Hildreth and D. Marr discovered that if an image were first filtered (smoothed) at different *spatial bandwidths,* and if the resulting images were subjected to differential edge detection, the resulting edge maps could be correlated to find the edges of real objects,[6, 7] which is to say that the true edges tend to be found

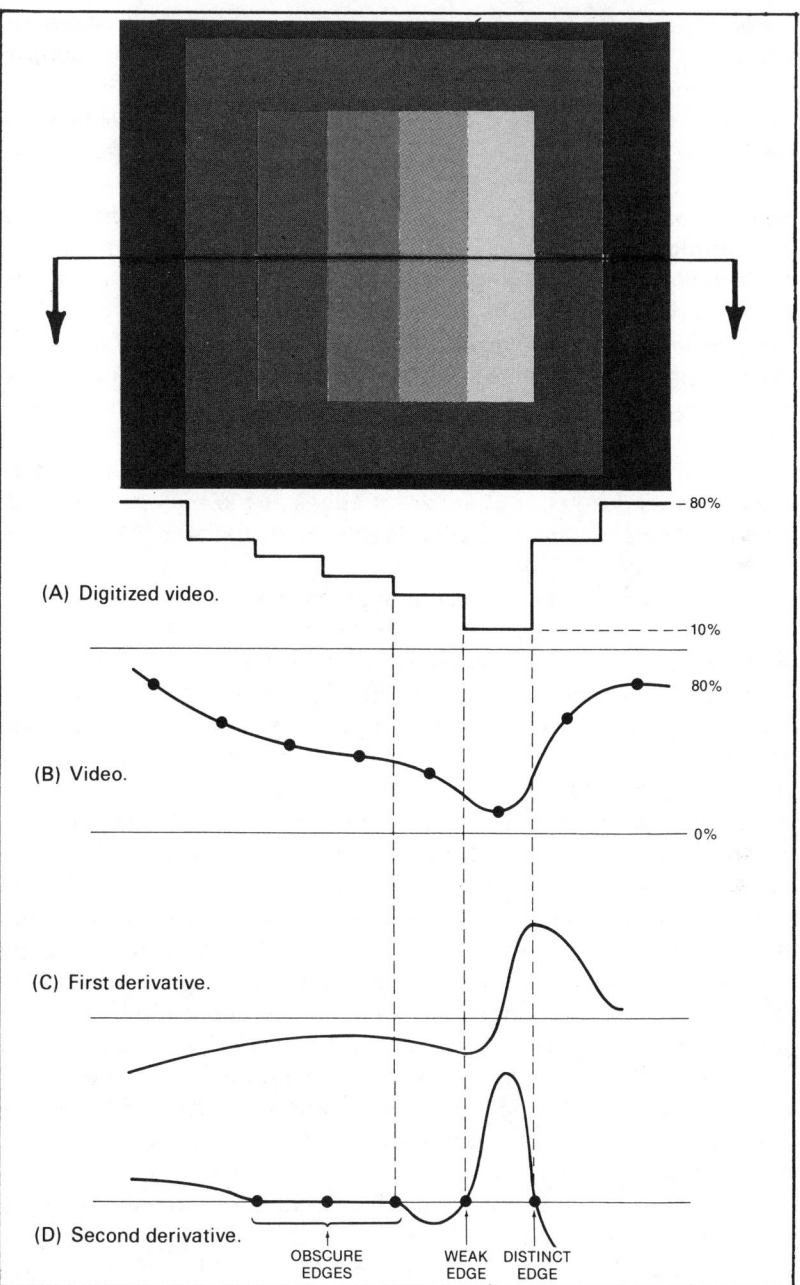

(A) Digitized video.

(B) Video.

(C) First derivative.

(D) Second derivative.

OBSCURE EDGES   WEAK EDGE   DISTINCT EDGE

Fig. 4-12. Video processing (simplified to one axis).

where the apparent edges (as determined at each resolution) coincide.

To understand this, Fig. 4-13 shows the same signal smoothed by simple averaging. As in the previous examples, the problem has been kept in one dimension. In Fig. 4-13A, the signal is not smoothed (which is the equivalent of smoothing by one pixel). In B the image is smoothed by averaging each pixel with the one to its left. This is something of a problem for a computer because the resulting value of the calculation should properly be stored at the left boundary of the pixel (halfway between the two pixels that were averaged), but this is not possible since each pixel represents a discrete address. For this reason, practical systems will usually average with odd-numbered spatial resolutions only (such as the example in column C that uses a spatial resolution of 3).

Finally, Fig.4-13C shows a filter resolution of three pixels. Notice the diminished number of apparent edges at the higher degrees of filtering. The edges detected in this column would tend to be those of larger, more global objects.

Filtering an image in two dimensions is not as difficult as might be expected. A simplified example of this is shown in Fig. 4-14. Each *filtered* pixel value is calculated by averaging nine pixel values representing the center pixel and its eight immediate neighbors. It becomes obvious that this is not quite right, since the center pixel should be weighted more heavily than its neighbors. Furthermore, if the second nearest neighbors were included in the filter calculation, they should be weighted more lightly than the immediate neighbors. The exact weighting algorithm is not so obvious; however, a Gaussian or similar profile is usually used. The reason for using a flat profile in this example is simply to allow the readers to rapidly verify their understanding of the process. Notice that the outer columns and rows are not calculated in Fig. 4-14B because each of these locations is missing at least one neighbor. In an actual application, the video processor would have to use some special rule for handling these boundary locations. This rule could simply restrict the boundaries of the filtered image (losing the border locations), or it could adjust the weighting of a particular set of neighbors depending on their number. If very large (low-resolution) filtering numbers were to be used, the second of these approaches might be dictated to prevent loss of a large area of the image.

The digital approximation of the first derivative of the filtered image is shown in Fig. 4-14C. Calculated by the application of the Pythagorean theorem, this approximation is a simple case of finding

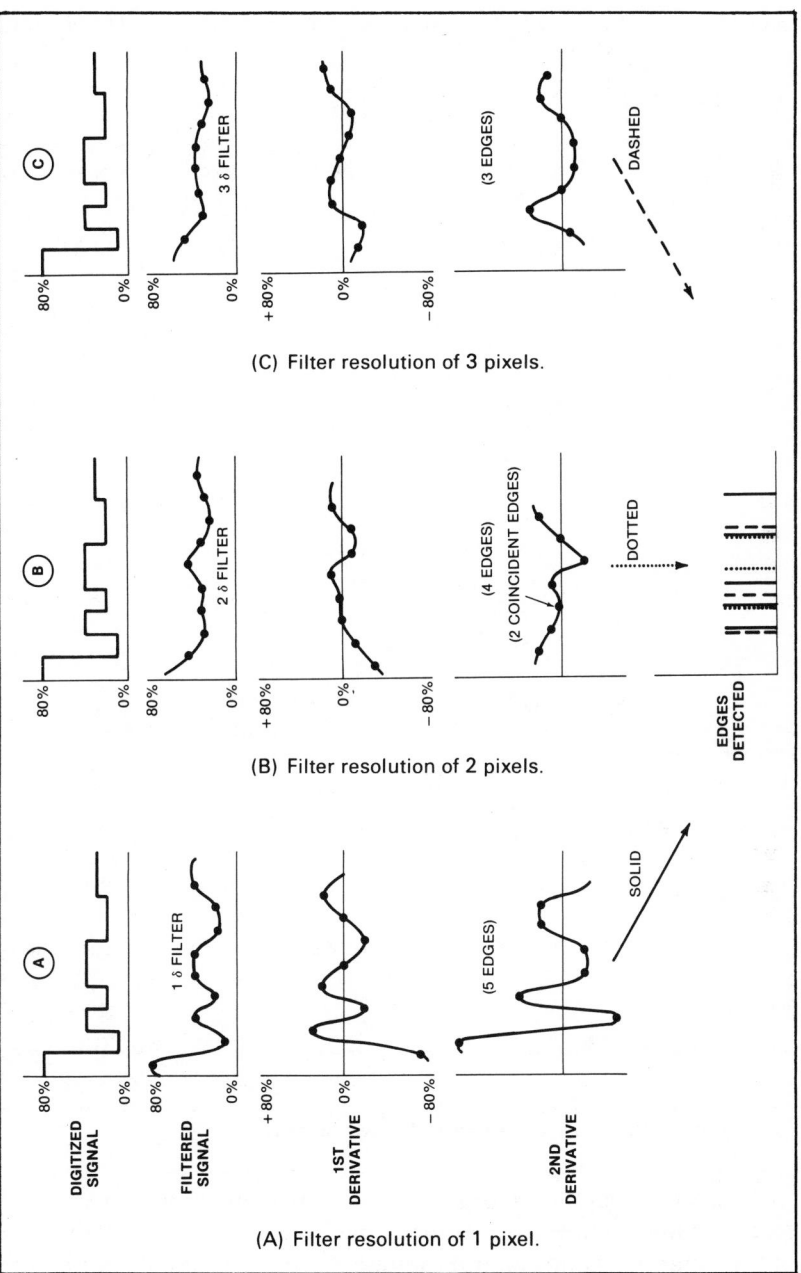

(C) Filter resolution of 3 pixels.

(B) Filter resolution of 2 pixels.

(A) Filter resolution of 1 pixel.

Fig. 4-13. Edge detection by comparison between filters.

197

UNPROCESSED VIDEO INTENSITY ARRAY

	A	B	C	D	E
1	60	80	70	80	90
2	70	60	50	60	80
3	70	40	30	40	70
4	90	50	10	50	50
5	80	40	0	40	30

(A) Unprocessed video intensity array.

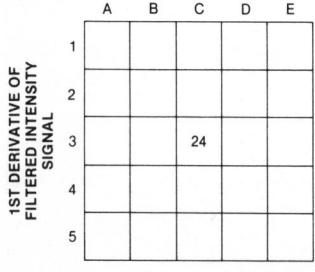

INTENSITY FILTERED BY UNWEIGHTED 1ST NEIGHBOR AVERAGING

	A	B	C	D	E
1					
2		59	57	63	
3		52	43	49	
4		46	33	36	
5					

$\Delta x = [3D] - [3B] = 49 - 52 = -3$
$\Delta y = [4C] - [2C] = 33 - 57 = -24$
$\Delta i = \sqrt{\Delta x^2 + \Delta y^2}$
$\Delta i = \sqrt{(-24)^2 + (-3)^2} = 24.1$

(B) Intensity filtered by unweighted first neighbor averaging.

1ST DERIVATIVE OF FILTERED INTENSITY SIGNAL

	A	B	C	D	E
1					
2					
3			24		
4					
5					

(C) First derivative of filtered intensity signal.

Fig. 4-14. A simplified example of image processing.

the unsigned magnitude of the result of two perpendicular vectors. Notice that the boundary conditions have once again caused the loss of the outer rows and columns, and in this case we are left with only the center pixel. Even though real images are much larger, the boundary losses are also larger, as large filter numbers are commonly used. This is one of the many places where trade-offs must be made. If the process is being done in hardware, the circuit complexity will be increased by the use of special boundary calculations. If software is used to perform the computation, it will be slowed by these additional boundary calculations. As always, the decision will depend on the system requirements. Notice that the second derivative (not shown) would be calculated from the first derivative in the same way that the first derivative was derived from the filtered-image array.

Optical Considerations in Edge Detection

In real-world applications, often many extraneous edges in the background may be distracting to the robot. These can sometimes be filtered out by the use of the optical characteristic known as *depth of field*. The depth of field is the range of distances from the lens over which the focus remains sharp. Opening the aperture tends to let in more light and at the same time it decreases the depth of field. If the robot knows the approximate range to the object, it can be made to adjust the aperture in such a way as to improve its recognition process. A robot with a fixed depth of field may suffer from the equivalent of nearsightedness or farsightedness, unless the aperture is *stopped-down* (closed) to a point that the depth of field is almost infinite. Since aperture control requires very little expense and since it does not burden the processor, it should be considered when vision problems include cluttered backgrounds. It should be remembered that the adjustment of the iris also changes the incident light on the camera target and that this change must be compensated by either electrical sensitivity adjustment and/or by the insertion and deletion of neutral-density optical filters.

Generally, larger lenses are more effective when a narrow depth of field is required. The actual figure of merit for a lens in this regard is the *f-number*. The f-number is defined as the focal length divided by the lens diameter. The smaller this number, the more light the lens collects and the narrower the depth of field.

At this point we have discussed several different methods by which the edges of an object can be determined. For the robot to understand how these edges relate to the real-world object or objects in its field of view is quite a separate problem although the means used to do this may be profoundly affected by the edge-detection method used.

Closing the Edge Contours

Objects that are entirely in the field of view of the robot are generally bounded by closed contours of edges. It is, therefore, natural to approach object recognition by attempting to determine which edges form closed shapes and discarding edge fragments. A simple set of rules for accomplishing this has been described by Wilf.[8] Wilf's algorithm describes the following steps:

1. Examine the image memory and check for the number of neighboring pixels that are *edge pixels*:
   a. If zero then store the pixel (by address) in a list of completed edges (it is an isolated point). Erase the point from the image edge map memory.
   b. If one, then store it in a list of end points (it terminates chain).
   c. If two, then ignore it (it is in the middle of a chain).
   d. If greater than two, then store it in a list of node points (it is at the junction of two or more edges.
2. Trace around each edge by starting at an end point or at a node point as found in the respective tables that were built during step 1. The eight neighbors are examined in a clockwise manner starting at the right-hand neighbor. When a neighbor is found that lies on an edge, it is added to the chain. If the new point is not a node point, it is erased from the image memory to avoid tracing it again. The neighbors of the new edge point are then examined counterclockwise, starting at the first pixel counterclockwise from the previous pixel and stopping at the one clockwise of it. In this way the edge is entirely erased except for node points. The process is repeated on node points until they have no neighbors, and then they are erased. By this point the image will contain only simple closed curves because the nodes contained only midpoints that were ignored by step 1. Curves intersecting with other curves or with line segments will have been destroyed. (Unfortunately, this destruction may occasionally eliminate desirable features.)
3. Examine the remaining image until one of the curves is detected. At this point the curve is followed as in step 2 (but there will be no nodes). As it is followed certain calculations may be made (see the next section) and it is erased from the image memory. Either all or some of the points of these curves may be stored in lists, depending on how many of the calcula-

tions are done as the curve is traced. In this way the remaining pixels of the image are erased, and the simple closed curves are reduced to lists of points and/or statistics. Some of the methods of processing these shapes will be discussed in more detail in the next section.

One of the disadvantages of this algorithm is that it tends to destroy any closed curve that is touched by another edge (Fig. 4-15A). For this reason, it may be desirable to modify step 2 in such a way that edges are removed by starting only at end points (and not node points). Under this modified technique, when a node point is encountered it is tested for neighbors, and if it has none, it is eliminated. If the node point has one or more neighbors, it is left in the map, and the process starts at another end point. This process may leave compound closed curves (as in Fig. 4-15B), which will be somewhat more ambiguous to identify than the simple closed curves of the original algorithm, but they more closely resemble the subject. If necessary, the object recognition algorithm may be used on each shape that is formed by the various combinations of overlapping closed curves, but this may overload the processor if the image becomes too complex.

The basic concept of determining the entire outline of an object is appealing, but it has obvious problems. For example, if a robot is to identify one part in a pile, it may not be able to find a single part whose outline is entirely unobscured. The larger an object, the less likely it is to have a perfect, closed outline that is not obscured by glare or shadows. So, it is often desirable to look for the smaller features of an object when trying to identify it. For example, the scissors of Fig. 4-10 might be identified by either of the finger holes or by the screw (not shown in the binary image). Once an object was located, it could be placed on a viewing surface, or otherwise manipulated, for closer examination. It is relatively easy to see that some of the more unique features of objects are not represented by closed curves. The points of the scissors, for example, offer unique shapes that might be of interest in identification. Other methods of identification use the mathematical characteristics of these features as will be discussed in the section on three-dimensional vision.

It is important to remember that the algorithms discussed thus far are only some of the more popular alternatives. Each has strong points and weak points. The robot system may require a processing speed that precludes performing these algorithms in software. The technique must then be simplified or more processing must be done in dedicated hardware. Unfortunately, the reaction of university researchers to the through-put problem is often not suitable to the

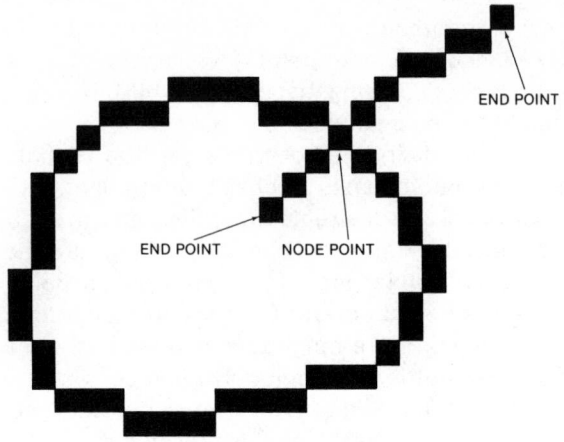

(A) Closed curve with node.

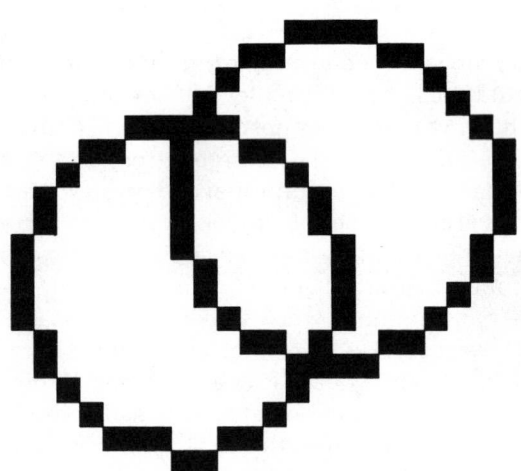

(B) Compound closed curves.

Fig. 4-15. Intersecting edges.

practical robot designer. Researchers are often prone to bring more and more powerful computers to bear on the problem, using large-scale parallel processors to improve performance. The result is usually prohibitively expensive for most applications. Fortunately, dedicated LSI and VLSI (Very Large Scale Integration) devices offer an economical solution. These will be developed as soon as a significant market demand becomes apparent to the semiconductor manufacturers. With this in mind, some development work is already concentrating on hardware augmentation rather than purely software implementation.[5]

TWO-DIMENSIONAL PROCESSING

Two-dimensional image processing has reached the point of being both reliable and practical enough for industrial applications. In general, practical applications of image processing are limited to well-illuminated, high-contrast images of items of interest to the robot (Fig.4-16). These subjects are typically confined to a plane, and the robot uses the vision system to determine the orientation or position of the subject or to measure some aspect of it. The outline of the object is usually obtained by one of the general methods discussed in the preceding section.

Courtesy Object Recognition Systems.

Fig. 4-16. A high-speed video inspection system.

Some of the applications are extremely simple. For example, a large robot arm might be used to perform several tasks on an automobile as it passed on an assembly line (see Chapter 3 concerning relative position control). In such an application the robot might be programmed to perform a task that required an accuracy beyond the tolerances of its position accuracy and the tolerances of the automobile. Inserting several small screws in holes would be typical of such a problem. While the normal position controls of the arm would cause the manipulator to track the approximate area of the holes, a camera located on the end of the arm would allow the robot to identify the exact location of the holes. This information would be used to correct the position of the manipulator by either adjusting the main servos, or by operating a small secondary positioner between the end of the larger arm and the manipulator (Fig. 4-17).

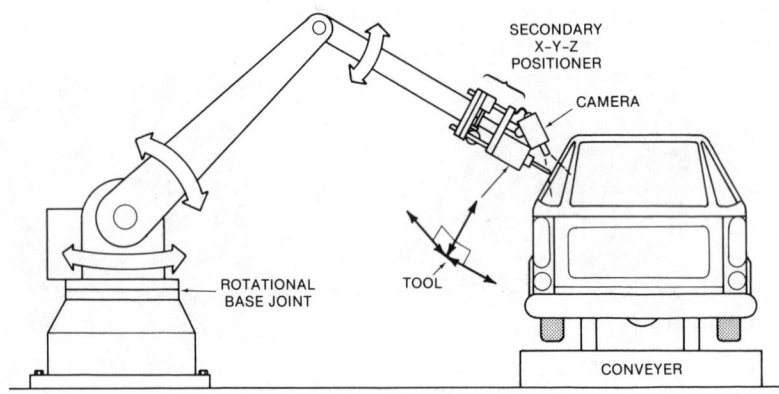

Fig. 4-17. Using vision to eliminate accumulated error.

Another application might require a robot to pick up a randomly oriented part from the conveyor belt and assemble it into a product. The robot might also do a cursory visual inspection on the object before picking it up. If the part failed to meet the requirements of the test, a conditional branch path of arm motion (see Chapter 2) would deliver the part to a scrap bin. In this example, the robot would have to study the outline of the component before picking it up (due to the random orientation). Because of this inspection, a comparison to acceptable limits might represent very little extra program complexity.

It is extremely important that the visual program of the robot be flexible to make training and retraining as simple and trouble free as possible. Typical training procedures require only that the robot be shown the part in a number of orientations. This training usually includes all of the stable orientations of the part when it is allowed to rest freely on a flat surface.

RECOGNIZING THE CLOSED-EDGE IMAGE

Assuming that the edge of an object (or some unique feature of an object) has been determined by a method such as those mentioned in the previous pages, there remains the problem of identifying the object and determining its orientation. In the following paragraphs, it will be assumed that the object is confined to a plane, and that the camera is pointed perpendicularly at the plane. The object may be rotated about an axis perpendicular to the plane, but the characteristics of the silhouette must be known to the vision system. Some symmetrical objects (such as a can) have a silhouette that does not vary with rotation about one or more of its axes. Thus, objects that have a limited number of profiles when in their stable positions may be allowed additional degrees of freedom, as long as the vision system is programmed with the characteristics of each stable profile. Several basic techniques for recognizing objects of this type will be discussed, along with their relative strengths and weaknesses.

Template Matching

Perhaps the simplest object recognition technique is template matching. This technique involves storing the entire image of the object in memory (usually in a binary form) for comparison to the images seen by the camera. As shown in Fig. 4-18, the processor simply finds the lowest vertical (Y) pixel (or pixels) location for each image and sets this as the position of the comparison X axis (Y = O). The same process is used to find the location of the comparison Y axis. The height and width of each image can then be determined by looking for the *highest* vertical and horizontal pixel position values as measured with respect to the comparison axes. If the height and width do not agree closely with those of the template, the process will either be terminated with a negative match result, or a scaling algorithm will be executed. This algorithm would modify the unknown image to have the correct height and width. In systems where range is constant, scaling is never necessary. It should be noted that scaling

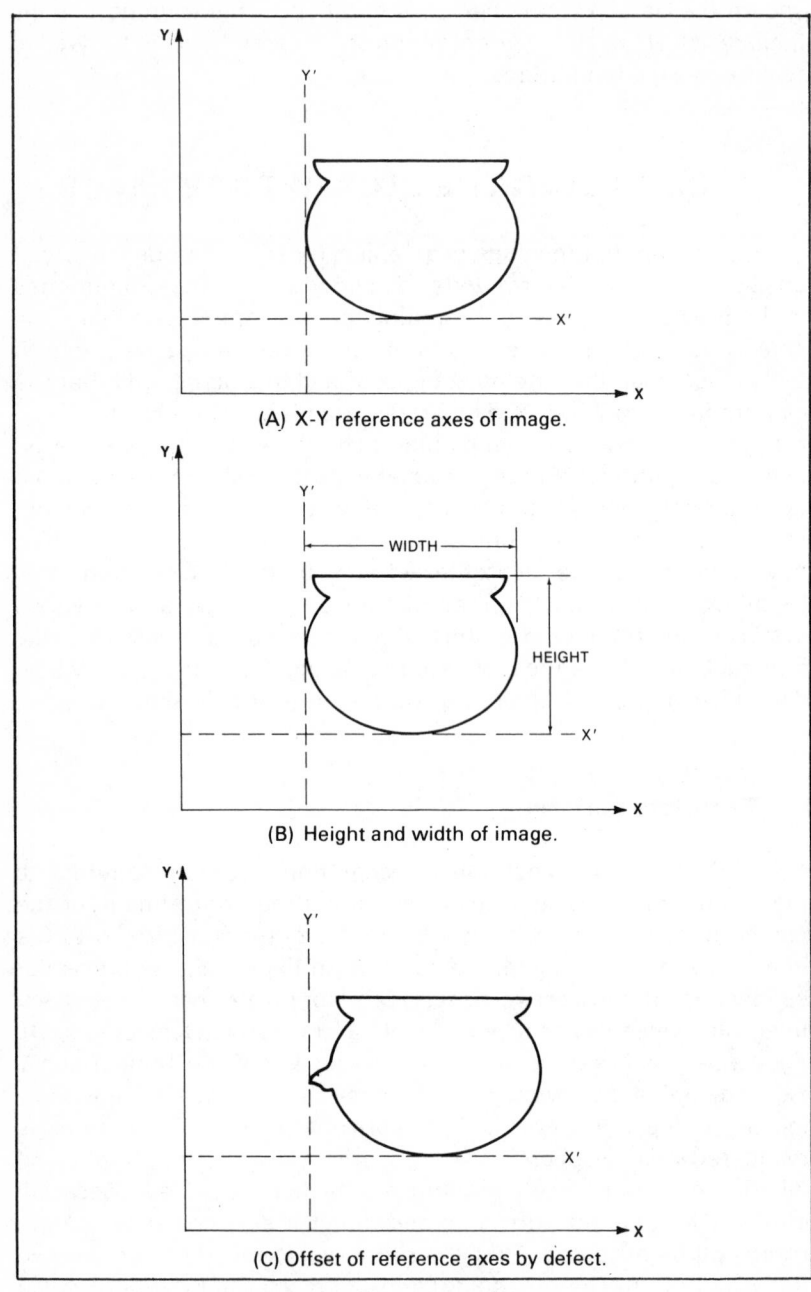

(A) X-Y reference axes of image.

(B) Height and width of image.

(C) Offset of reference axes by defect.

Fig. 4-18. Setting reference axes for template matching.

requires very high resolution images if unacceptable distortion is to be avoided.

Once the height and width dimensions do agree (within predetermined tolerances), the two images are compared on a pixel-by-pixel basis. If the ratio of matching pixels to nonmatching pixels is sufficiently high, the algorithm returns a positive match result.

If the image orientation of the object is not known, the template matching program may contain a rotational algorithm that would slowly rotate the unknown image until correct height and width measurements were found (assuming a fixed range/scale factor). At each rotational position where these dimensions matched, the pixel-by-pixel comparison would be repeated until either a match were found, or all possible orientations had been tested. This rotational process requires a considerable amount of processing and cannot be combined with scaling algorithm discussed earlier.

In addition to the disadvantages already mentioned, this technique has the additional weakness of being extremely sensitive to slight irregularities along the reference (bottom and left) edges as shown in Fig. 4-18C. This weakness is because such irregularities can cause an offset of the reference axes that will result in a disproportionate number of pixel mismatches.

In applications such as conveyor systems, template matching may be quite useful since the range and orientation are usually constant and processing can be accomplished rapidly. Unfortunately, robots generally do not have this advantage, and thus the process is of very limited use. Although many improvements can be made to the general process described above, template matching is not inherently flexible enough to be considered in more depth here.

Advanced Methods of Image Recognition

Several characteristics of a profile can be used to identify it. Proper selection and processing of these features will produce an algorithm that is relatively tolerant of orientation and range variations, and of small surface irregularities.

One of the most obvious features of the profile is the length of the border, which can be determined by tracing the edge. During tracing, a running total is kept of the length. If the next pixel on the curve is located directly above or below the present pixel, or directly to the right or left, a value of one is added to the total. If, however, the next pixel is at a diagonal position (45°) from the present pixel, it represents a longer segment. In this case, a value of 1.414 is added (Pythagoras to the rescue again). The border length is dependent on the distance from the camera to the object, but not on the rotation (although some rotational error will creep in due to quantitizing effects).

AREA, MOMENTS, AND THE AXIS OF MINIMUM INERTIA

As in other chapters, the bulk of the equations associated with mathematics of moments has been carefully concealed in Appendix A. The reason for this is twofold. The first reason is to provide easy reference, and the second reason is to avoid "equation shock" on the part of the reader. It is not necessary to understand these equations to understand the concepts behind them. In fact, once the concepts are mastered, the equations can be easily understood.

To begin with, moments are nothing more than the accumulative effect of the *grains* of area of the silhouette with respect to their distance from an axis (or axes). In pure mathematics, these grains are infinitely small, and the process of accumulating their effect is done through the use of integration. With video processing, there is no advantage in considering a finer resolution than the pixel, and the accumulative process is easier to visualize. Equations 4-2A–F represent a mathematical technique for tracing around the border and determining the moments.

The contribution of each pixel to the particular moment is determined by its distance from the axis about which the moment is being calculated. If the first moment is to be calculated, this implies that each pixel will make a contribution with respect to the *first* power of its distance from the axis. Thus each pixel would simply be multiplied by the distance from the axis for the first moment, but it would be multiplied by the square (second power) of the distance for the second moment. A simple three-pixel shape is shown in Fig. 4-19A. The first three moments are calculated for this shape. Notice that the zero moment is found by multiplying each pixel area (1) by the zero power of its distance from the axis. Since any number taken to the zero power is 1, each pixel contributes to this moment without respect to its distance from the axis. The result is that the zero moment is familiar to us as the area of the shape.

In mathematical notation, the first moment with respect to the X axis would be $m(1,0)$. The second moment with respect to the Y axis would be $m(0,2)$. It should be noted that a moment can be calculated

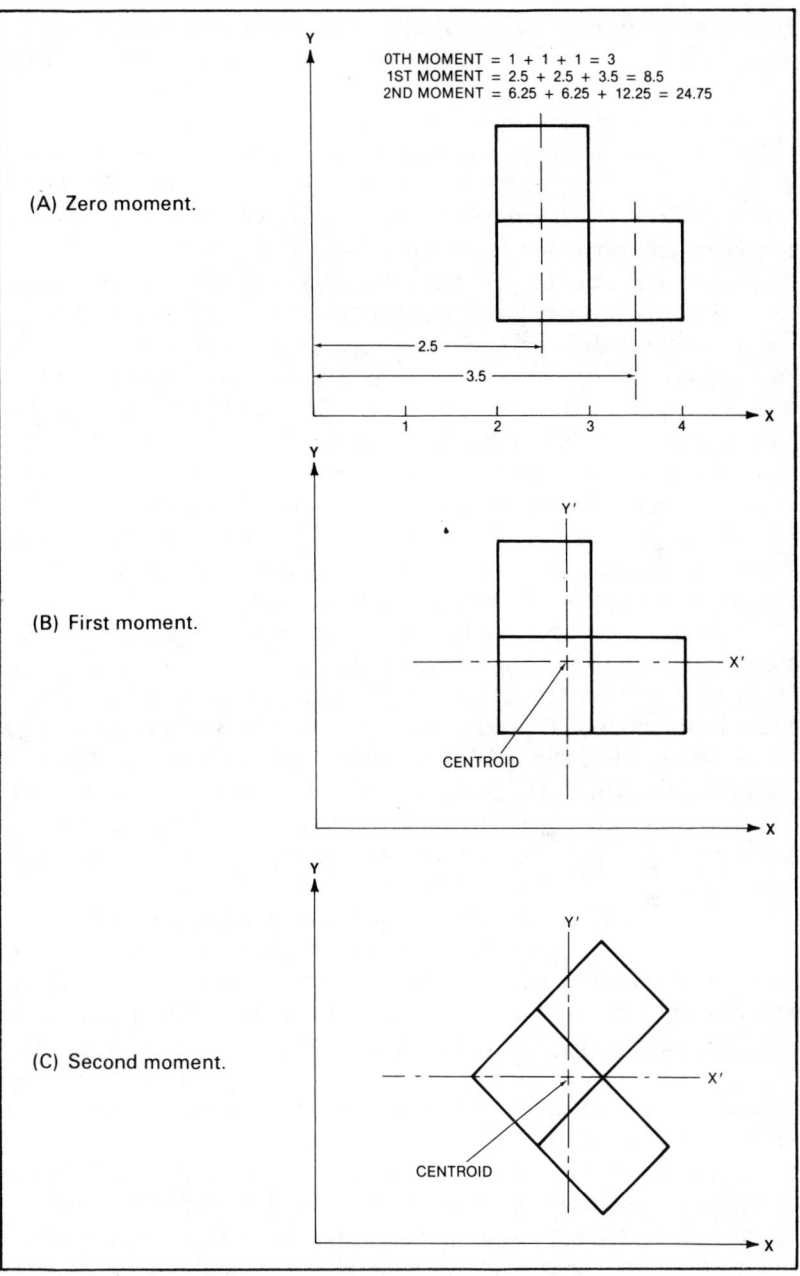

(A) Zero moment.

0TH MOMENT = 1 + 1 + 1 = 3
1ST MOMENT = 2.5 + 2.5 + 3.5 = 8.5
2ND MOMENT = 6.25 + 6.25 + 12.25 = 24.75

(B) First moment.

(C) Second moment.

Fig. 4-19. Locating the central axes.

209

with respect to both axis. For example, m(1,2) would be the first moment with respect to the X axis, and the second moment with respect to the Y axis.

Notice also that the first and second moments are calculated by taking the distance from the axis to the *center* of the pixel. This is because the middle is the center of mass of the pixel. For those readers who are interested in the mysteries of Appendix A, this factor shows up directly in the last term of Equation 4-2C.

The first moment is familiar from physics as the static balancing force. Some readers might prefer to think of this as the "leverage" of the shape about the axis. In Fig. 4-19B, the Y axis has been moved to a point where the first moment is zero. This is the *balance axis* and it runs through the centroid (balance point) of the shape. A second axis has been added to 90° to the first, and it too has been placed so that the first moment with respect to it is zero. The place where these two axes cross is the centroid or center of gravity of the shape. *No matter what the orientation of the shape might be with respect to our axes, the intersection of the axes of zero first moment will always occur at the same place (the centroid) on the silhouette.*

It is not necessary to add the contribution of each pixel to find the areas and moments. The process can be done while stepping around the border of the shape, and this is the principle on which Equations 4-2A–F are based. To show how simple this is, Equation 4-1 (a combination of Equations 4-2A and B) has been allowed out of exile in Appendix A and has been placed in Fig. 4-20. Notice that the equation calls for the summation of the links from L(link) = 1 to n, where n is the last link. In Fig. 4-20, the links have been given letters a through i, so the summation will actually be a to i.

To verify that this process works, we can trace the links of the simple chain of Fig. 4-20, as shown in the accompanying table. The result of 6.5 units can be verified by simply adding the enclosed squares. Notice that the calculation is done in a counterclockwise direction starting with link a and ending with link i. The calculation for each link is done at the counterclockwise termination of the link. If the process is done clockwise, the result will be the same but negative in value.

It would be very nice if some mathematical quality could give us an indication of the orientation of the object. The axis of least inertia (second moment) is just such a quality. If we were to attempt to twirl a shape about an axis running through its centroid at various angles, the orientation that provided the least inertia would be the axis of least inertia (obviously), and the angle of this axis with respect to the system X axis is usually referred to as the angle of least inertia. It can be proven that this axis will always run through the centroid, and thus it will always correspond to one of the axes of zero first moment. The axis of least inertia of the silhouette of a can, for example, would be a line drawn through the center of the can from end to end. Some shapes, like circles, contain more than one axis of least inertia, but typically the profile program can recognize this condition and make an arbitrary value choice. Quantizing error will often do this automatically.

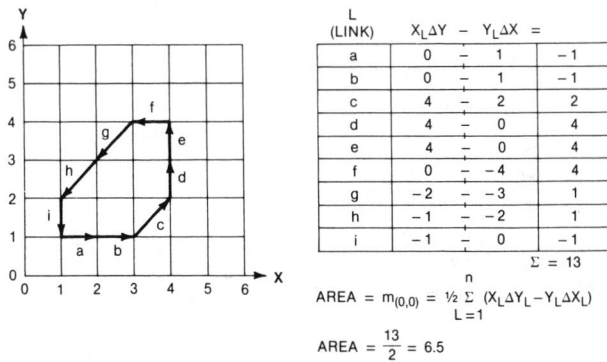

Fig. 4-20. Calculation of the area of a closed curve.

It turns out, that if one takes a particular shape and defines a set of axes running through the centroid such that one axis (say the Y axis) is the axis of minimum inertia, the various moments about these axes tend to be *unique to the shape.* Thus, these values make a simple and elegant *signature* for the shape. The process for finding the signature values on an unknown shape could be as follows (see Fig. 4-21):

1. Calculate the first moments about the arbitrary system X and Y axes, and calculate the area of the shape.
2. Use the values calculated in step 1 to find the centroid, and then find the axis of least inertia. This eliminates the effects of position and rotation.
3. Calculate the various moments about the axis of least inertia (Y') and the corresponding X' axis. These would be called the *moments with respect to the axis of least inertia.*
4. Normalize the moments calculated in step 3, by dividing by the area as calculated in step 1. This eliminates the effec· of range

211

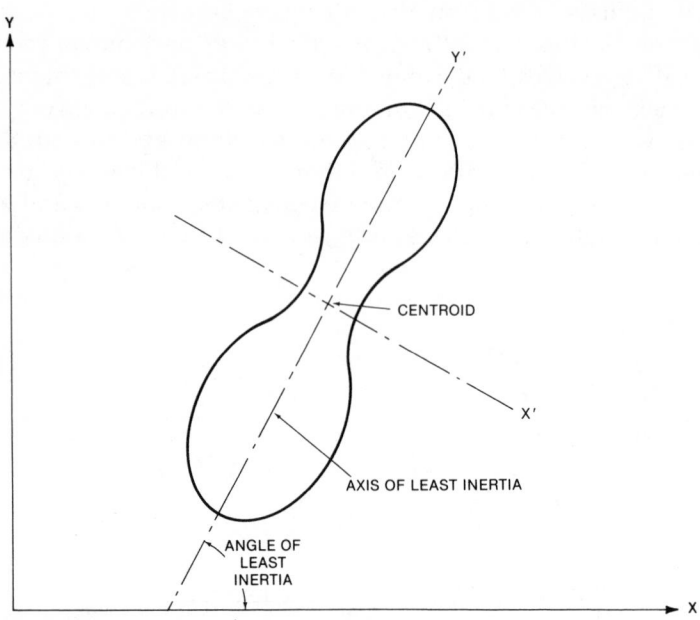

Fig. 4-21. Finding the axis of least inertia.

(size) variations, and the resulting values are referred to as the *normalized moments about the axis of least inertia.*

5. Place these values in a table, and compare them to the same parameters for known objects.

THE SRI ALGORITHM AND MOMENT INVARIANTS

In practice, it is not necessary to find the axis of least inertia.[10] The fact that it does exist and that the normalized moments about it tend to be unique is sufficient. By combining the equations for calculating the axis of least inertia with those for calculating the moments, certain quantities can be calculated that are independent of the rotation of the object. These are called the moment invariants,[8, 9] and they may be found while tracing the border of the shape as was done with the area. The equations for calculating these invariants are given in Appendix A, Equations 4-2 through 4-7. The process is done by the following steps:

212

1. Find the moments about the arbitrary system axes (Equations 4-2). This is the only step that requires border tracing.
2. Find a new set of axes, parallel to the first, but running through the centroid. These are called central axes, and can be found by simply dividing the first moments calculated in step 1 by the area (Equations 4-3A and B). This eliminates the effect of position.
3. Calculate the moments about these new axes from the moments found in step 1, and the axis position values (with respect to the old axis) as were found in step 2. These new moments are referred to as the central moments because they are calculated with respect to the central axis (Equations 4-5A–E).
4. Find the *normalized central moments* of the shape by taking the area into account (Equation 4-6). This step eliminates the effect of size (range).
5. Calculate the moment invariants from the normalized central moments (Equations 4-7A–D). This eliminates the effect of rotation.
6. Place the invariants in a table, and compare them with the moment invariants of known objects.

Although these equations may look rather impressive, they are in fact rather simple to accomplish with a computer. Even a microprocessor can do an adequate job of image recognition using moment invariants, as long as there are not too many extraneous shapes to test, and provided there is not a critical time constraint. Where speed must be improved, several processors may be placed in tandem, with each doing a different function (i.e., camera interfacing, filtering, edge detection, and moment calculations). This kind of processing is sometimes called *pipelining*. The first processor would be working on the most recent frame, while each successive processor would be processing data from progressively older frames.

COLOR VISION

The addition of color to a vision system can greatly improve the ability of the system to recognize objects because color is a feature that is independent or range, position, and rotational effects. Color images are usually stored in very much the same way as gray-scale images, but the image data is placed in three memory fields corresponding to the three primary additive colors (red, green, and blue). The use of color imaging will not be given much treatment here since it can be handled in much the same way as already described for gray-scale information. The color derivatives for determining edges will be somewhat more difficult to derive, but they will also be much less prone to error. In addition to this advantage, color displays can provide more information about the subject (Fig. 4-22).[M7]

THREE-DIMENSIONAL VIDEO PROCESSING AND RANGE DETERMINATION

While the methods described in the preceding pages are most applicable to the fixed pick-and-place manufacturing robot, it is inev-

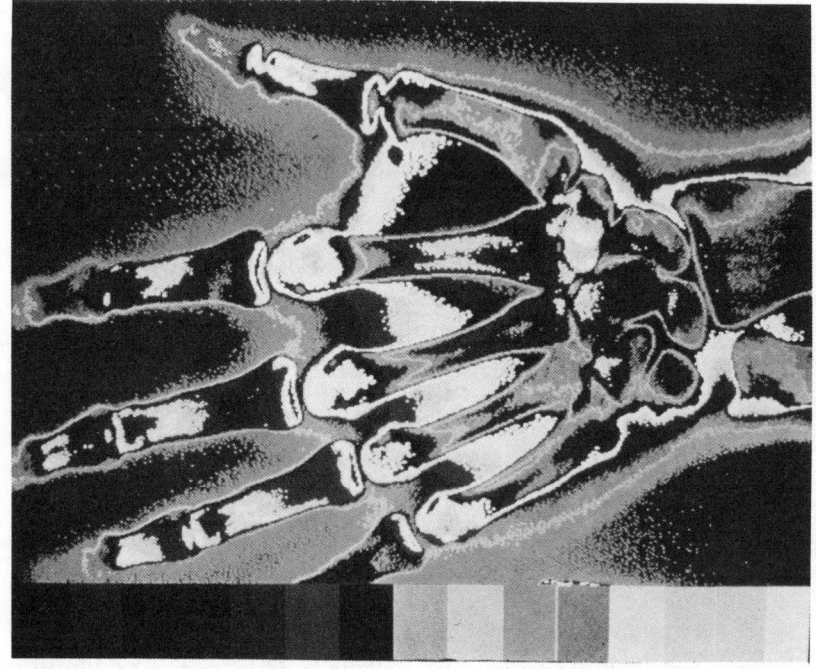

Courtesy Digital Graphic Systems, Inc.

Fig. 4-22. A high-resolution (480 × 512 × 4) color image as digitized by the Digital Graphic Systems, Inc. Model CAT 400.

itable that mobile robots will begin to make a serious commercial appearance in the near future. At least one manufacturing robot has already been equipped with a standard PUMA arm and has demonstrated the ability to move about within a limited work area performing tasks.

The first significant success at navigation in an unknown environment by *vision alone* probably belongs to Dr. Hans P. Moravec.[1] His inauspicious looking "Rover" robotic cart (developed as part of his doctoral thesis at Stanford University) was able to navigate through a maze of real and fabricated obstacles in order to reach a predetermined location (Fig. 4-23A). Although the processing time requirements caused the cart to move very slowly through the maze, it did so with impressive accuracy on indoor courses. Outdoors, the cart did not fare quite so well since glare and the play of shadows caused scenes to change rather significantly between images. This problem should not distract from the importance of the success since it would be largely cured with increased processing speed.

One of the several novel innovations of this project was the abandonment of binocular range finding that had been the traditional approach used by the predecessors of "Rover." Based on the human method of range finding, the binocular method had been found to be very error prone in most robotic applications. This problem was caused by the processor tending to match a feature in one view with a similar but wrong feature in the other view. To realize how easily this could happen, imagine a robot looking at a picket fence. Since the slats are almost identical, the system would have to be very comprehensive to reliably avoid mismatching slats in two views. This is further aggravated when the features are small interest operators (features selected by the processor because they are mathematically interesting).

Faced with this problem, Moravec made an observation of nature that provided a solution. When observing captive lizards preying on insects, he was struck by the accuracy of their sense of range. The lizards would always gauge their leaps with an impressive accuracy. This accuracy was all the more interesting to Moravec in light of the fact that they had opposing eyes that could not possibly be used for binocular range determination. He then noticed that a lizard would sway its head slowly before attacking. Moravec realized that this continuous motion would allow a single-image camera system to determine range. Most importantly, objects whose ranges were to be determined could be tracked as the camera moved, thus eliminating false corrections.

(A) Navigating an obstacle course at Stanford's Artificial Intelligence Laboratory.

Courtesy Dr. Hans P. Moravec.

(B) Showing the slide mounted camera.

Fig. 4-23. Moravec's cart.

216

Moravec's camera (Fig. 4-23B) was mounted on a 50 cm long slide, and digitized the image nine times as it traversed the slide. A program called the *interest operator* was executed on the first of the nine images. This program would locate a number of mathematically unique features in the picture. This program was biased toward locating at least some reference points in every area of the image. A *correlator* program was then executed on the remaining eight images of the sequence. This program would identify the reference points chosen by the interest operator. A *camera solver* problem was then used to determine the range of each feature from its displacement in the various images. The feature *map* developed by the camera solver program was then used by a *navigator program* to move the cart forward a short distance in such a way as to eventually reach a predetermined destination (without hitting any vertically displaced features). The software was sufficiently sophisticated to ignore even the most striking feature if it was laying flat on the ground and thus did not represent an impediment to the cart.

After each step forward by the cart, a version of the correlator program would attempt to locate as many of the features of the previous ranging sequence as possible. From the apparent displacement of these references, the motion of the cart was deduced, and the process was repeated. It is interesting to realize that, although the cart might appear to be avoiding a desk, it might actually be avoiding a handle, or the corner of one of the drawers, or a combination of such features.

The interest operator located features that were relatively unique in the reference image in order to minimize the chance of the correlator mismatching them in other views. Features with sharp corners or very small closed shapes (such as spots) have this characteristic. The interest operator checked overlapping windows of 4 × 4 pixels (Fig. 4-24), by determining the variance in each of four directions. This was accomplished by determining the sums of the squares of the difference between adjacent pixels in each of the directions. The minimum (worst) of these four values was taken as the figure of merit for the sub-window. The sub-window with the highest figure of merit in each region of twenty-five overlapping sub-windows was taken as the local interest feature.

The correlator used a convolution process for matching features in various images. To minimize the chance of incorrect identification of features, the convolution process matched low resolution, and then successively higher resolution images in the two views. This was accomplished by defining a search area in the view in which the

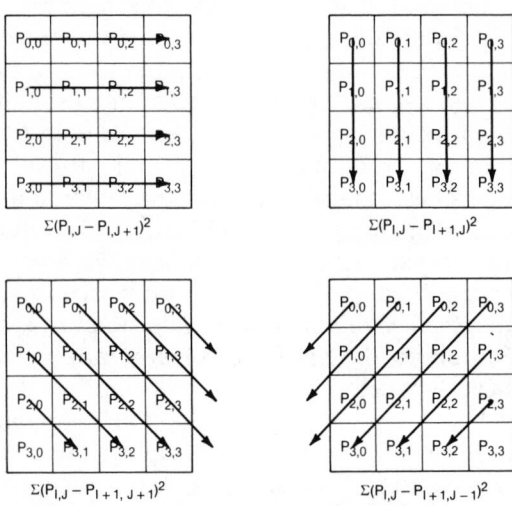

$\Sigma(P_{I,J} - P_{I,J+1})^2$

$\Sigma(P_{I,J} - P_{I+1,J})^2$

$\Sigma(P_{I,J} - P_{I+1, J+1})^2$

$\Sigma(P_{I,J} - P_{I+1,J-1})^2$

Courtesy Dr. Hans P. Moravec.

Fig. 4-24. Moravec's method of determining interesting features from local intensity variance.

point was to be located. This search area was centered around the suspected position (taken from the position in the reference image). The resolution of this area was reduced by averaging (as described in the earlier section on the variable resolution digitizer). A small patch of the reference image around the reference point was then filtered to the same scale and resolution as the search area. This patch was then convolved with all of the possible locations within the search area, and the best match was taken as the location. The process was then repeated with successively smaller, higher-resolution areas (Fig. 4-25). Not only was the reliability of the matches improved over convolving once at full resolution, but the process was faster because of the smaller number of total convolutions required.

While this process was ideally suited for Moravec's application, it has the disadvantage of not tolerating image rotation, illumination, or size (range) variations. The sensitivity to illumination changes was at least partially responsible for the problems that the cart experienced outdoors. In applications where range, illumination, and rotation are fixed, this technique may be used to supplement (or replace) the techniques mentioned earlier.

While the earlier edge-detection techniques may have application in future three-dimensional processing work, the object recogni-

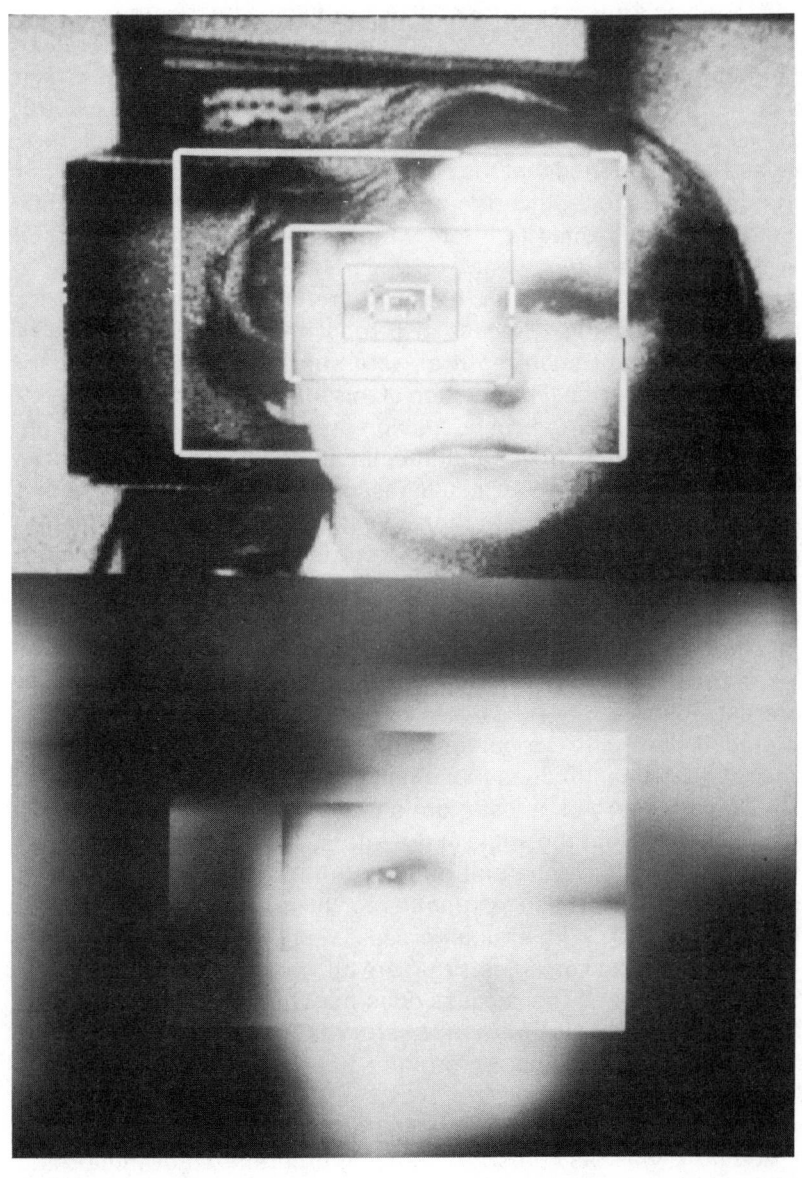

Fig. 4-25. Variable-resolution windowing.

tion processes (using moment invariants) would be of use only when the robot was searching for a known object.

Moravec's interest operator also poses some interesting possibilities with respect to the types of problems discussed earlier. Recalling the problem of finding a pair of scissors in a pile of other scissors, it was mentioned that a vision recognition system would have a better chance of finding a finger hole than of finding a whole pair of scissors. It was also mentioned that the points of the scissors presented rather unique silhouettes. Since the points are not bounded shapes, they could not be subjected to the moment invariant process discussed earlier. These features would, however, have been most exciting to Moravec's interest operator. It may, therefore, be advantageous for the image "learning" process of a robot to include the storing of information about such features as a supplement to the moment invariant information. The addition of this information could make the overall operation of the robot's vision system more *robust* (reliable over a variety of adverse conditions). Failing to find an identifiable bounded shape, the robot could resort to looking for the interest operators. The results might then give the robot enough information to cause it to either give up the effort or to attempt to rearrange the objects present for possible identification of bounded shapes. In order to use the interest operator for more precise identification, it would have to be improved. The weakness of Moravec's interest operator lies in the fact that it makes measurements only at 45° angles. An interest point may yield largely different values as the object is rotated. Although this problem could be improved,[1] it is only one of several related techniques that could be explored.

Some research has been done on techniques that take advantage of the fact that the angle of the intersection of two straight lines is a constant with range and rotation (in the plane of confinement discussed earlier). These techniques can be used with the edge maps discussed earlier. With enough processing power, these angles can even be recognized on objects that are allowed additional degrees of rotational freedom. The process does not work well on curved surfaces, however, and is thus most useful when dealing with Euclidean shapes (shapes bounded by straight lines such as most artificial structures).

One additional improvement made by Moravec that will be mentioned here was his camera linearizing program. To overcome the vidicon's distortion (mentioned at the beginning of this chapter), the cart's camera was calibrated at the beginning of each run by showing it a pattern of precisely spaced spots. Two polynomials (X and Y) were derived for spatial correction of images. These polynomials were used in the obstacle avoider to correct for camera roll, tilt, focal length, and long-term geometry distortion.

STRUCTURED LIGHT VISION SYSTEMS

Besides binocular ranging and Moravec's sliding camera ranging technique, it is also possible to determine range by the use of a camera and a *structured* light source such as a laser. Present structured light systems use single "swept" laser beams, arrays of modulated laser beams, sheets of noncoherent light, and holograms (Fig. 4-26). This is an extremely interesting and promising new field, but it is beyond the scope of this book.

THE VISION SYSTEM AS AN INTEGRATED PART OF THE ROBOT

As important as the quality of the vision process is the quality of its interface to the other programs and sensors of the robot. As the sophistication of robots increases, this factor will become even more important. A mobile robot with an elaborate vision system will have to be capable of interpreting the data from this system in context with other information. For example, visual reference points could be lost in an image due to a flash of light. This loss could be offset by information from a simple ultrasonic ranging system. It is therefore important that the three-dimensional models developed by these programs be compatible in format.

CONCLUSIONS

Those who are new to this subject may be very disappointed that I have not provided a simple and concise answer to their individual requirements. However, there are no such answers. The number of design trade-offs increases with the complexity of a system, and vision systems are far from simple. Until more comparative results are available from the various techniques mentioned in this chapter (and some that were not mentioned), the designer will have to make some tough choices. Even when the state of this technology has evolved, these choices will still be dependent on the tasks expected of the robot.

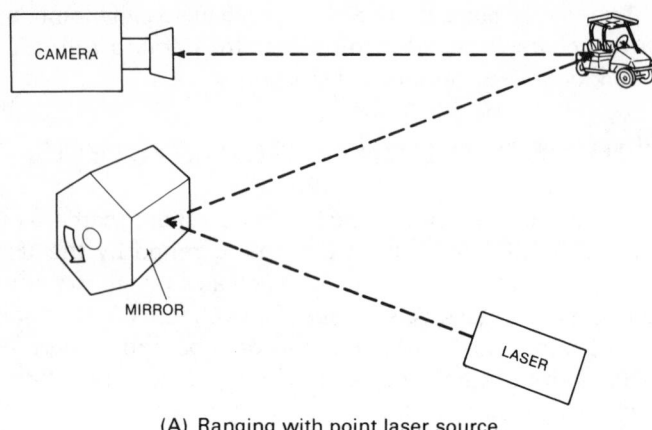

(A) Ranging with point laser source.

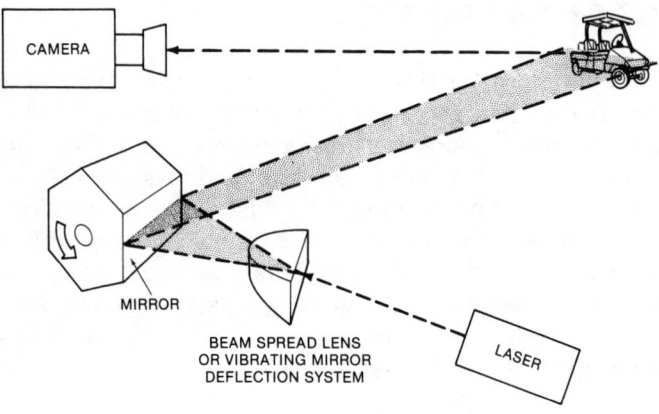

(B) Ranging with a line laser source.

Fig. 4-26. Ranging with laser sources.

REFERENCES

1. Moravec, Hans P. *Obstacle Avoidance and Navigation in the Real World by Seeing Robot Rover.* Computer Science Dept., Stanford University, 1980.
2. *Reference Data for Radio Engineers.* Howard W. Sams & Co., Inc., Indianapolis, IN, 1975.

3. "Video Signal Input." *Robotics Age,* March/April 1981.
4. Weinstein, Martin. *Android Design.* Hayden Book Company, Inc., Rochelle Park, NJ, 1981.
5. Hildreth, Helen C. "Edge Detection in Man and Machine." *Robotics Age,* September/October 1981.
6. Marr, D. VISION. W.H. Freeman Co., San Francisco, CA, 1981.
7. Hildreth, Helen C. *Implementation of a Theory of Edge Detection.* MIT TR-579.
8. Wilf, Joel M. "Chain Code." *Robotics Age,* March/April 1981.
9. Hu, M. *Visual Pattern Recognition by Moment Invariants.* IRE Transactions on Information Theory, IT-8, 1968.
10. Wong, R. and Hall, E. "Scene Matching With Invariant Moments." *Computer Graphics and Image Processing,* Vol. 8, 1978.

MANUFACTURER REFERENCES

M1. MC 520 CCD Camera
E.G.& G. Reticon 345 Potrero Avenue
Sunnyvale, CA 94086
(408) 738-4266

M2. Dithertizer II™
Computer Stations Inc.
11610 Page Service Drive
St. Louis, MO 63141

M3. Plumbicon
Amperex Electronics
Providence Pike
Slatersville, RI 02876
(401) 762-3800

M4. Vidicon, Ultracon, Newvicon
RCA
New Holland Pike
Lancaster, PA 17604
(717) 397-7661

M5. Type 511, Type 611 Image Digitizer
Periphicon
P.O. Box 324
Beaverton, OR 97075
(503) 222-4966

M6. Video Digitizer 270A, Video Frame Store 274(C),
Programmable Digital I/O Module 720,
DMA Digital I/O 721
Colorado Video Inc.
Box 928
Boulder, CO 80306
(303) 940-3248

M7. CAT Color Imaging Boards
Digital Graphic Systems, Inc.
407 California Avenue, No. 1
Palo Alto, CA 94306
(415) 856-2500

M8. Image Processing Systems
Octek Inc.
7 Corporate Place, South Bedford Street
Burlington, MA 01803
(617) 273-0851

Mathematical Equations

MOTOR EQUATIONS FOR CHAPTER 1[*]

---

*Courtesy Kollmorgen Corp., Inland Motors Div.

Motor Voltage and Current

	Symbol	Unit
Armature current	I	A
Armature resistance	R_{mhot}	$1.4 \times 1.125 \times R_m$
Brush drop	BD	$V(= 0 \text{ or } 2)$
Developed torque	T_{dev}	lb-ft
Motor back emf	K_B	V/krpm
Motor torque sensitivity	K_T	lb-ft/A
Output torque	T_{output}	lb-ft
Speed	N	rpm
Static friction	T_F	lb-ft
Terminal voltage	V_T	V
Torque losses in motor	T_{loss}	lb-ft
Viscous damping loss	F_I	lb-ft/krpm

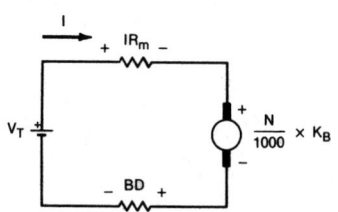

1) $T_{dev} = T_{output} + T_{loss}$

2) $T_{dev} = I \times K_T$

3) $T_{loss} = (N/1000) \times F_I + T_F$

4) $V_T = IR_{mhot} + (N/1000) \times K_B + BD$

5) $T_{load} = T_{output}$

6) from 1) ... $T_{output} = T_{dev} - T_{loss}$

7) from 5), 6) ... $T_{load} = T_{dev} - T_{loss}$

8) from 2), 3), 7) ... $T_{load} = I \times K_T - ((N/1000) \times F_I) - T_F$

9) from 8) ... $I \times K_T = T_{load} + (N/1000) \times F_I + T_F$

10) from 9) ... $I = (T_{load} + (N/1000) \times F_I + T_F)/K_T$

Motor Resistance and Brush Drop

The motor winding resistance, R_m, is rated at 25°C armature temperature and has a tolerance of ±12.5%. This resistance increases by about 40% at maximum armature temperature to give a "worst case" resistance value of $R_{mhot} \cong 1.4 \times 1.125 \times R_m \cong 1.575\, R_m$, Ω.

For many motors, the resistance of the brushes is included in R_m. For cases where R_m does not include brush resistance, a value of 2 volts may be used as the voltage drop across the brushes at rated continuous current.

Motor Efficiency

This procedure will determine the efficiency of the motor at any specific speed and torque.

Motor EFF $= \eta$

Choose a point
on the PC curve.

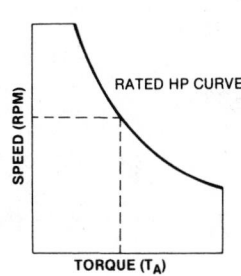

Determine watts out:

$$\text{watts out} = (746)(HP)$$

Power in $= EI$

$$I = \frac{T_A + T_F + (N/1000)\ (F_I)}{K_T}$$

$$E = IR_m\,(1.52) + K_B\,(N) + BD$$

$$P_{IN} = EI = \text{watts in}$$

$$\eta = \frac{\text{watts out}}{\text{watts in}}$$

$$HP = \frac{N \times T}{5252}$$

where,

N is in rpm,
T is in lb-ft.

Need to know:

1) HP_{cont}

2) T_F

3) F_I

4) K_T

5) R_m

6) K_B

7) BD

where,

HP_{cont} is continuous horsepower,

T_F is motor static friction (lb-ft),

F_I is motor viscous damping (lb-ft/1000 rpm),

K_T is torque constant (lb-ft/amp),

R_m is motor armature resistance (Ω),

K_B is voltage constant (volts/rpm),

BD is brush drop $= 2$ volts.

228

Types of Acceleration Profiles

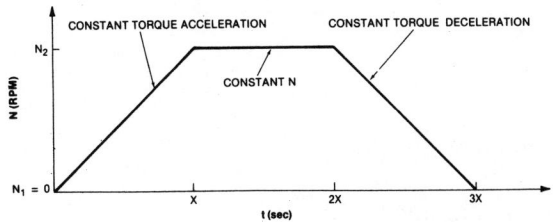

(A) One-step profile (one of the most efficient practical profiles.)

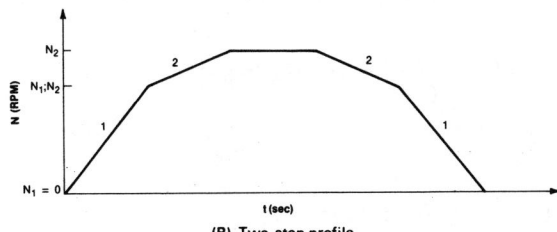

(B) Two-step profile.

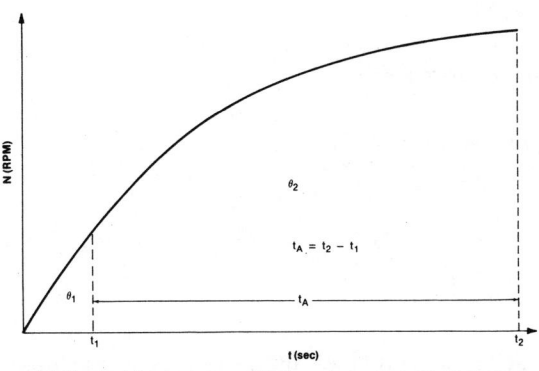

(C) Constant horsepower profile from t_1 to t_2, constant torque acceleration from 0 to t_1.

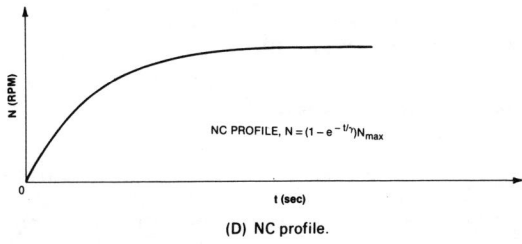

(D) NC profile.

Constant Torque Acceleration, One-Step or Two-Step

On summing torques, one gets

$$T_A - T_d = T\alpha$$

where,

T_A is acceleration torque (lb-ft),

T_d is deceleration torque (lb-ft),

$T\alpha$ is $J_T\alpha$, acceleration torque for a frictionless system (lb-ft).

$$J_T = \frac{J_L}{n^2 \eta_n} + J_m \text{ (lb-ft-sec}^2)$$

where,

n is gear ratio, motor revs/load revs,

η_n is gear box efficiency,

J_L is load inertia (lb-ft-sec^2),

J_m is motor inertia (lb-ft-sec^2).

$$\alpha = \frac{n(N_2 - N_1)}{t_a} \quad \frac{\pi}{30} \text{ (rad/sec}^2)$$

where,

N_2 is high load speed (rpm) (see figures for accel profiles),

N_1 is low speed (rpm) (see figures for accel profiles),

t_a is time to reach N_2 from N_1 (sec),

t_d is time to reach N_1 from N_2 (sec).

$$T_{trav} = \frac{FT}{n \times \eta_n} + T_F + n \frac{N_2 + N_1}{2} \frac{F_l}{1000} \left(\begin{array}{l} \text{load and friction} \\ \text{torques reflected to} \\ \text{the motor shaft} \end{array} \right)$$

230

The third term of T_{trav} is an average motor "viscous damping" torque. The definition of all other terms may be found in the list of application parameters. Substitutions of the definitions of the terms and rearrangements of the torque summation equation yield:

$$T_A = J_T \alpha + T_{trav}$$

or,

$$T_A = \frac{J_T n (N_2 - N_1) \pi}{30 \, t_a} + T_{trav} \text{ (lb-ft)}$$

$$T_d = \frac{T_T n (N_2 - N_1) \pi}{30 \, t_d} - T_{trav} \text{ (lb-ft)}$$

Thus, during acceleration:

$$t_a = \frac{J_T n (N_2 - N_1) \pi}{30 (T_A - T_{trav})} \text{ (sec)}$$

and during deceleration:

$$T_d = \frac{J_T n (N_2 - N_1) \pi}{30 (T_A + T_{trav})} \text{ (sec)}$$

Also,

$$HP = \frac{T_A n N}{5252} \text{ (hp) (peak torque during acceleration)}$$

$$T_{rms} = T_A \text{ (lb-ft) (root-mean-square torque)}$$

and

$$\theta = \frac{(N_2 + N_1)}{2} \; \frac{\pi}{30} \, t \text{ (rad), (angular load displacement in time t)}$$

One-Step Acceleration Profile and Thermal Efficiency

One-step acceleration profiles are among the most efficient (i.e.) having minimum motor temperature) practical profiles. It has been shown that the most efficient one-step (i.e., trapezoidal) profile involves a 1/3-1/3-1/3 scheme: acceleration during one-third of the cycle period, constant velocity during one-third of the period and deceleration during the last third of the period.

(Reference: Kuo, B.C. and J. Tal (eds.), *Incremental Motion Control, Volume I: DC Motors and Control Systems.* SRL Publishing Company, Champaign, Illinois, 1978., Chapter 15).

Horsepower Acceleration

ACCELERATION

$$T_A = \frac{J_T}{T_{trav}} \left[\frac{n\pi}{30} (N_1 - N_2) + \frac{550\,HP}{T_{trav}} \ln\left(\frac{(5252\,HP/T_{trav}) - nN_1}{(5252\,HP/T_{trav}) - nN_2}\right) \right] (sec)$$

$$\theta_2 = \frac{550\,HP}{T_{trav}} \times t_A - \frac{J_T}{2 \times T_{trav}} \times (N_2{}^2 - N_1{}^2) \left(\frac{n\pi}{30}\right)^2 \quad (rad)$$

$$T_{rms} = \left[\frac{550\,HP \times J_T}{T_A} \ln\left(\frac{(5252\,HP/nN_1) - T_{trav}}{(5252\,HP/nN_2) - T_{trav}}\right) \right]^{1/2} \quad (lb\text{-}ft)$$

$$T_A = \frac{5252\,HP}{nN} - T_{trav} \ (lb\text{-}ft)$$

where,

HP is given constant power value (hp),

θ_2 is acceleration or deceleration associated with the velocity N_2, (rad/sec^2). All other variables have been defined under "constant torque acceleration."

DECELERATION

$$T_d = \frac{J_T}{T_{trav}} \left[\frac{n\pi}{30} (N_2 - N_1) + \frac{550\,HP}{T_{trav}} \ln\left(\frac{(5252\,HP/T_{trav}) + nN_1}{(5252\,HP/T_{trav}) + nN_2}\right) \right] (sec)$$

$$\theta_2 = - \frac{550\,HP}{T_{trav}} \times t_d + \frac{J_T}{2 \times T_{trav}} \times (N_2{}^2 - N_1{}^2) \left(\frac{n\pi}{30}\right)^2 \quad (rad)$$

$$T_{rms} = \left[\frac{550\,HP \times J_T}{T_d} \ln\left(\frac{(5252\,HP/nN_1) + T_{trav}}{(5252\,HP/nN_2) + T_{trav}}\right) \right]^{1/2} \quad (lb\text{-}ft)$$

$$T_d = - \frac{5252\,HP}{nN} - T_{trav} \ (lb\text{-}ft)$$

NC Acceleration

Numerically controlled acceleration uses the following velocity equation:

$$N(t) = N_F (1 - e^{-t/\tau})$$

where

$N(t)$ is speed in rpm (as a function of time),

N_F is final speed,

τ is time constant ($= 0.06/\text{NC gain}$).

Peak torque needed:

$$T_p = \frac{J_T \pi N_F}{30\tau} + T_{trav} \qquad \text{(lb-ft)}$$

Peak horsepower needed:

$$HP_p = \frac{T_{trav}^2 \tau 30}{J_T \pi (21{,}008)} + \frac{T_{trav} N_F}{10504} + \frac{J_T \pi N_F^2}{120\tau(5252)} \qquad \text{(hp)}$$

Time to accelerate to 95% of final speed $= 3\tau$ \qquad (sec)

Effective torque over 3τ:

$$T_{EFF}|_{3\tau} =$$

$$\sqrt{\frac{1}{3\tau} \left[\frac{(.99752)J_T^2 \pi^2 N_F^2}{(2)\,30^2 \tau} + \frac{(.95021)J_T \pi N_F T_{trav}}{15} + T_{trav}^2(3\tau) \right]} \quad \text{(lb-ft)}$$

Effective Horsepower over 3τ:

$$\frac{(J_T^2 \pi^2 N_F^4 - 2J_T \pi N_F^3 30\tau T_{trav} + 30^2 \tau^2 T_{trav}^2 N_F^2)}{30^2 \tau (5252)^2} \left(\frac{.99752}{2} \right) = A$$

$$\frac{(-2J_T^2 \pi^2 N_F^4 + 2J_T \pi 30\tau T_{trav} N_F^3)}{30^2 \tau (5252)^2} \left(\frac{.99988}{3} \right) = B$$

$$\frac{(2J_T \pi N_F^3 30\tau T_{trav} - 30^2 \tau^2 T_{trav}^2 N_F^2)}{30^2 \tau (5252)^2} \quad (.95) = C$$

$$\frac{J_T{}^2\pi^2N_F{}^4}{30^2\tau(5252)^2}\left(\frac{.99999}{4}\right) = D$$

$$\frac{-\,2J_T\pi N_F{}^3\,30\tau T_{trav}}{30^2\tau(5252)^2}\left(\frac{.99752}{2}\right) = E$$

$$\frac{(3)\,30^2\tau^3 T_{trav}{}^2 N_F{}^2}{30^2\tau^2(5252)^2} = F$$

$$HP_{EFF}|_{3\tau} = \sqrt{\frac{1}{3\tau}\,(A+B+C+D+E+F)} \qquad (hp)$$

Highest NC Gain for a Given System

From the total inertia of a system, traverse torque, horsepower rating of the motor, and final speed, the lowest time constant (highest NC gain) can be found for this system using the following equations:

For NC accel profile:

$$N(t) = N_F(1 - e^{-t/\tau}) \qquad (rpm)$$

Lowest τ available:

$$2\tau = \frac{4J_T\pi HP(5252) - 2J_T\pi N_F T_{trav}}{30\,T_{trav}{}^2}$$

$$-\sqrt{\frac{4J_T\pi HP(5252) - 2J_T\pi N_F T_{trav}}{30\,T_{trav}{}^2} - \left(\frac{2J_T\pi N_F}{30\,T_{trav}}\right)^2}$$

NC gain (highest) $= 0.06/\tau$ (inches/min/mil)

Note: Peak torque should not be exceeded on initial slope.

Check to see if T_p is exceeded using:

$$T_p = \frac{J_T\pi N_F}{30\tau} + T_{trav} \quad (lb\text{-}ft)$$

where,

   J_T is total inertia of the system (lb-ft-sec^2),
   T_{trav} is traverse torque including all system frictions (lb-ft),
   HP is HP rating of motor (hp),
   T_p is peak torque of motor (lb-ft),
   N_F is final speed reached (rpm),
   τ is time constant (sec).

GEAR AND ACTUATOR EQUATIONS FOR CHAPTER 1*

Inertia of a Uniform Steel Rod

The inertia of a cylindrical object is given by:

$$J = \tfrac{1}{2}MR^2$$

where,

   M is mass in slugs,
   R is radius in feet.

$$M = \frac{weight}{g} \quad (g = 32.174 \ ft/sec^2)$$

$$weight = \pi r^2 l \delta$$

where,

   l is length in inches,
   δ is density in lbs/in^3,
   r is radius in inches.

Substituting,

$$J = \tfrac{1}{2} \frac{(\pi r^2 l \delta) \times R^2}{g} \ (slug \ ft^2)$$

---

*Courtesy Kollmorgen Corp., Inland Motors Div.

Converting R^2 to $\dfrac{r^2}{144}$,

$$J = \tfrac{1}{2}\ \frac{\pi r^2 l \delta r^2}{g(144)}\ \text{(lb-ft-sec}^2\text{)} \quad \text{(slug ft}^2 = \text{lb-ft-sec}^2\text{)}$$

Converting R^2 to $\dfrac{d^2}{4}$,

$$J = \frac{\tfrac{1}{2}\pi d^2 l \delta d^2}{4g\ (144 \times 4)}\ \text{(lb-ft-sec}^2\text{)}$$

where,

d = diameter in inches.

Collecting terms,

$$J = \frac{\pi \delta d^4 l}{4608g}\ \text{(lb-ft-sec}^2\text{)}$$

For steel,

$$T = \frac{\pi\ 490 d^4 l}{(12)^3\ 4608g}\ \text{(lb-ft-sec}^2\text{)}$$

$$J = \frac{490\ \pi d^4 l}{7962624g}\ \text{(lb-ft-sec}^2\text{)}$$

$$\delta = \frac{490}{(12)^3}\ \text{(lbs/in}^3\text{)}$$

Converting Linear Inertia to Rotational Inertia

Equate the kinetic energy of the linear and rotational inertia systems:

Linear:Rotational

$$\frac{1}{2}\ mV^2 = \frac{1}{2}\ J\omega^2$$

where,

ω is in rad/sec,
V is in ft/sec,

236

m is mass in slugs,

J is inertia in slug ft^2 (slug ft^2 = lb-ft-sec^2)

$$J = m \left(\frac{V^2}{\omega^2} \right)$$

$$V = S/t$$
$$\omega = \theta/t$$

where,

S is in feet,

θ is in radians.

$$J = \frac{\text{weight}}{g} \left(\frac{V}{\omega} \right)^2 \quad \text{(lb-ft-sec}^2\text{)}$$

where,

g = 32.174 ft/sec^2.

Substituting for V, ω,

$$J = \frac{\text{weight}}{g} \left(\frac{S/t}{\theta/t} \right)^2 \quad \text{(lb-ft-sec}^2\text{)}$$

Simplifying,

$$J = \frac{\text{weight}}{g} \left(\frac{S}{\theta} \right)^2$$

$$\frac{S}{\theta} \text{ in } \frac{\text{ft}}{\text{rad}} = \frac{\text{lead}}{2\pi \times 12}$$

where,

lead is in inches/revolution.

Substituting,

$$J = \frac{\text{weight}}{g} \left(\frac{\text{lead}}{24\pi} \right)^2 \quad \text{(lb-ft-sec}^2\text{)}$$

Simplifying,

$$J = \frac{\text{weight} \times \text{lead}^2}{g \times 576 \times \pi^2} \text{ (lb-ft-sec}^2\text{)} \cong \frac{\text{weight} \times \text{lead}^2}{1.82 \times 10^5} \text{ (lb-ft-sec}^2\text{)}$$

Reflecting Inertia Across a Speed Reducer

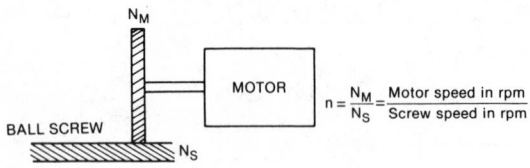

$$\text{Kinetic Energy} = \frac{1}{2}J\omega^2$$

Conservation of Energy

At Screw	At Motor

$$\frac{1}{2}J_{LS}\omega_S^2 = \frac{1}{2}J_{LM}\omega_M^2$$

where,

J_{LS} is load inertia at ballscrew,
J_{LM} is load inertia reflected to motor,
ω_S is speed of ballscrew in rad/s,
ω_M is speed of motor in rad/s.

Solving for J_{LM},

$$J_{LM} = J_{LS}\left(\frac{\omega_S^2}{\omega_M^2}\right)$$

Substituting,

$$J_{LM} = J_{LS}\left(\frac{N_S}{N_M}\right)^2 \quad \text{where } \frac{N_S}{N_M} = \frac{\omega_S}{\omega_M}$$

$$J_{LM} = J_{LS}\left(\frac{1}{n}\right)^2 \quad \text{where } \frac{1}{n} = \frac{N_S}{N_M}$$

$$J_{LM} = \frac{J_{LS}}{n^2}$$

Including inefficiencies of gearbox (η_n),

$$J_{LM} = \frac{J_{LS}}{n^2 \times \eta_n}$$

Note:

The reflected inertia is independent of the gearbox efficiency. Thus, the inclusion of η_n in the denominator of the expression for J_{LM} is physically meaningless but is a valid means of incorporating the torque frictional losses in the gearbox.

Reflecting Torque Across a Speed Reducer

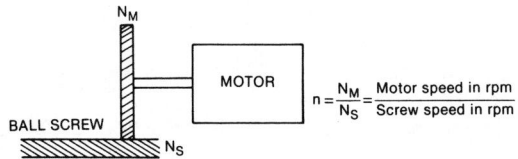

Given that HP is constant across a speed reducer,

$$\text{horsepower} = \frac{N \times T}{5252}$$

$$\text{HP at screw} = \frac{N_S \times T_{LS}}{5252}$$

$$\text{HP at motor} = \frac{N_M \times T_{LM}}{5252}$$

where,

T_{LS} is load torque at ballscrew,
T_{LM} is load torque at motor.

Since HP is constant,

$$\text{HP at screw} = \text{HP at motor}$$

$$\frac{N_S \times T_{LS}}{5252} = \frac{N_M \times T_{LM}}{5252}$$

Solve for T_{LM},

$$T_{LM} = \frac{N_S \times T_{LS}}{N_M}$$

where,

$$\frac{N_S}{N_M} = \frac{1}{n}$$

Substituting,

$$T_{LM} = \frac{T_{LS}}{n}$$

Including inefficiencies of the gear box (η_n),

$$T_{LM} = \frac{T_{LS}}{n \times \eta_n}$$

Speed: Linear to Rotational; Rotational to Rotational

$$\text{lead} = \frac{\text{inches}}{\text{revolution}}$$

$$\text{linear speed}/\text{lead} = \text{rotational speed (N)}$$

where,
   linear speed is in ipm,
   N is in rpm.

For a motor driving a load through a speed reducer,

$$n = \text{gear ratio} = \frac{N_m}{N_l} = \frac{\text{motor speed, rpm}}{\text{load speed, rpm}}$$

$$N_m = N_l \times n$$

MOBILITY EQUATIONS FOR CHAPTER 3

Equation 3-1. Finding the Static Moment Distance d' from the Robot's Relative Distance d and Pitch Angle ϕ

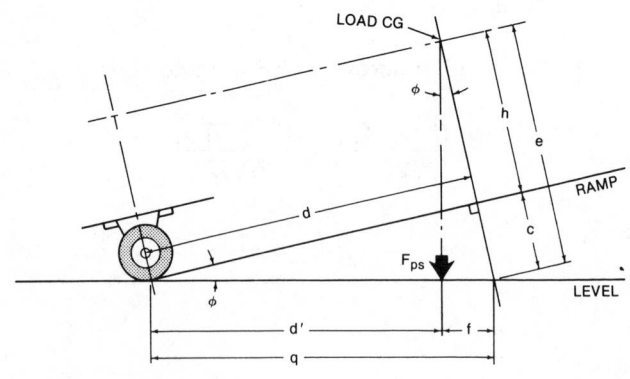

Note: Shown for positive pitch (ϕ).

Rotary Inertia Conversion Table

To Convert from A to B, multiply by entry in Table.

A \ B	gm-cm²	oz-in²	gm-cm-sec²	kg-cm²	lb-in²	oz-in-sec²	lb-ft²	kg-cm-sec²	lb-in-sec²	lb-ft-sec² or slug-ft²
gm-cm²	1	5.46745×10^{-3}	1.01972×10^{-3}	10^{-3}	3.41716×10^{-4}	1.41612×10^{-5}	2.37303×10^{-6}	1.01972×10^{-6}	8.85073×10^{-7}	7.37561×10^{-8}
oz-in²	182.901	1	0.186507	0.182901	0.0625	2.59009×10^{-3}	4.34028×10^{-4}	1.86507×10^{-4}	1.61880×10^{-4}	1.34900×10^{-5}
gm-cm-sec²	980.665	5.36174	1	0.980665	0.335109	1.38874×10^{-2}	2.32714×10^{-3}	10^{-3}	8.67960×10^{-4}	7.23300×10^{-5}
kg-cm²	1000	5.46745	1.01972	1	0.341716	1.41612×10^{-2}	2.37303×10^{-3}	1.01972×10^{-3}	8.85073×10^{-4}	7.37561×10^{-5}
lb-in²	2.92641×10^{3}	16	2.98411	2.92641	1	4.14414×10^{-2}	6.94444×10^{-3}	2.98411×10^{-3}	2.59009×10^{-3}	2.15840×10^{-4}
oz-in-sec²	7.06157×10^{4}	386.088	72.0079	70.6157	24.1305	1	0.167573	7.20079×10^{-2}	6.25×10^{-2}	5.20833×10^{-3}
lb-ft²	4.21403×10^{5}	2304	429.711	421.403	144	5.96756	1	0.429711	0.372972	3.10810×10^{-2}
kg-cm-sec²	9.80665×10^{5}	5.36174×10^{3}	1000	980.665	335.109	13.8874	2.32714	1	0.867960	7.23300×10^{-2}
lb-in-sec²	1.12985×10^{6}	3.17740×10^{3}	1.15213×10^{3}	1.12985×10^{3}	386.088	16	2.68117	1.15213	1	8.33333×10^{-2}
lb-ft-sec² or slug-ft²	1.35582×10^{7}	7.41289×10^{4}	1.38255×10^{4}	1.35582×10^{4}	4.63305×10^{3}	192	32.1740	13.8255	12	1

Torque Conversion Table

To Convert from A to B, multiply by entry in Table.

A \ B	dyne-cm	gm-cm	oz-in	kg-cm	lb-in	newton-m	lb-ft	kg-m
dyne-cm	1	1.01972×10^{-3}	1.41612×10^{-5}	1.01972×10^{-6}	8.85073×10^{-7}	10^{-7}	7.37561×10^{-8}	1.01972×10^{-8}
gm-cm	980.665	1	1.38874×10^{-2}	10^{-3}	8.67960×10^{-4}	9.80665×10^{-5}	7.23300×10^{-5}	10^{-5}
oz-in	7.06157×10^{4}	72.0079	1	7.20079×10^{-2}	6.25×10^{-2}	7.06157×10^{-3}	5.20833×10^{-3}	7.20079×10^{-4}
kg-cm	9.80665×10^{5}	1000	13.8874	1	0.867960	9.80665×10^{-2}	7.23300×10^{-2}	10^{-2}
lb-in	1.12985×10^{6}	1.15213×10^{3}	16	1.15213	1	0.112985	8.33333×10^{-2}	1.15213×10^{-2}
newton-m	10^{7}	1.01972×10^{4}	141.612	10.1972	8.85073	1	0.737561	0.101972
lb-ft	1.35582×10^{7}	1.38255×10^{4}	192	13.8255	12	1.35582	1	0.138255
kg-m	9.80665×10^{7}	10^{5}	1.38874×10^{3}	100	86.7960	9.80665	7.23300	1

$$d' = q - f$$

where,

$$q = \frac{d}{\cos \phi}$$

$$f = e \times \sin \phi = (h + c) \sin \phi$$

since,

$$c = d \times \tan \phi \rightarrow f = (h + d \times \tan \phi) \sin \phi$$

therefore,

$$d' = \frac{d}{\cos \phi} - (h \times \sin \phi) - (d \times \tan \phi \times \sin \phi)$$

Notice that the last term is very small at angles less than 15°. (@ 15°→tan 15° × sin 15° = .067)

Equation 3-2 (A, B, & C). Resolving the Gravitational Force Vector (Step 1)

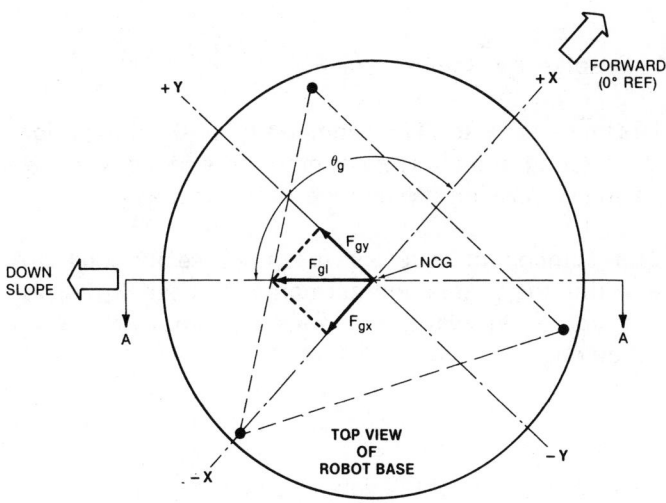

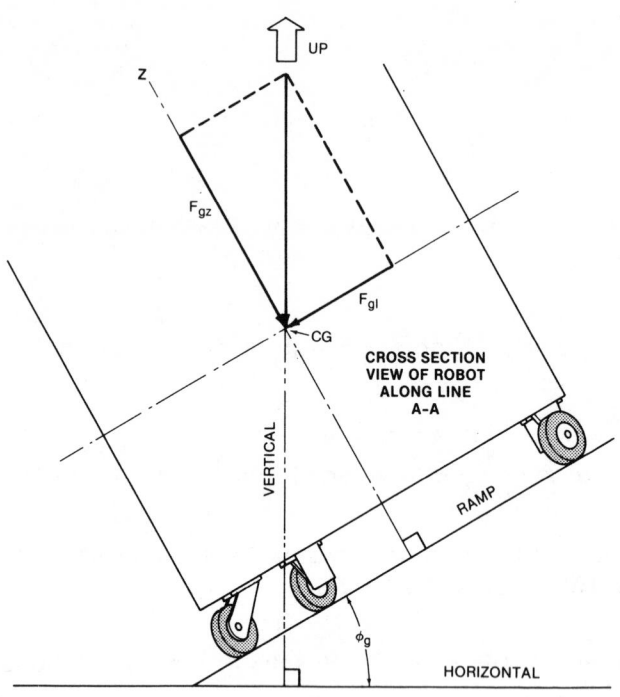

Step 1. Resolve the Gravitational Force.

First determine the lateral (F_{gl}) component of the gravitational force (F_g) that is acting in a plane parallel to the surface that the robot is running on and passing through the center of gravity (CG).

To find this component we resolve the gravity vector in plane A-A. We must also find the component normal to the base (F_{gz}) as this is the stabilizing force. We will then resolve F_{gl} into F_{gx} and F_{gy} for later vector addition to the dynamic forces.

Where,

θ_g is the azimuth angle of the down slope direction,
ϕ_g is the "tilt" angle of the grade,
F_g is gravitational force $= W =$ weight (units of lbs or newtons).

244

$$F_{gl} = F_g (\sin \phi_g) \qquad F_{gz} = F_g (\cos \phi_g)$$

or,

$$F_{gl} = W(\sin \phi_g) \qquad F_{gz} = W(\cos \phi_g) \qquad \text{(A)}$$

Further resolving F_{gl},

$$F_{gx} = F_{gl} (\cos \theta_g)$$

or,

$$F_{gx} = W(\sin \phi_g)(\cos \theta_g) \qquad \text{(B)}$$

Likewise,

$$F_{gy} = F_{gl} (\sin \theta_g)$$

or,

$$F_{gy} = W(\sin \phi_g)(\sin \theta_g) \qquad \text{(C)}$$

Equation 3-2 (D, E, F & G) Finding the Total Lateral Force F_t, the Deflection Azimuth θ_t, and the Deflection Distance d (Step 2)

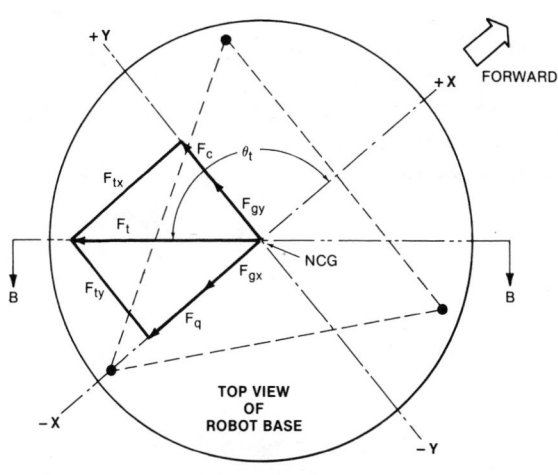

245

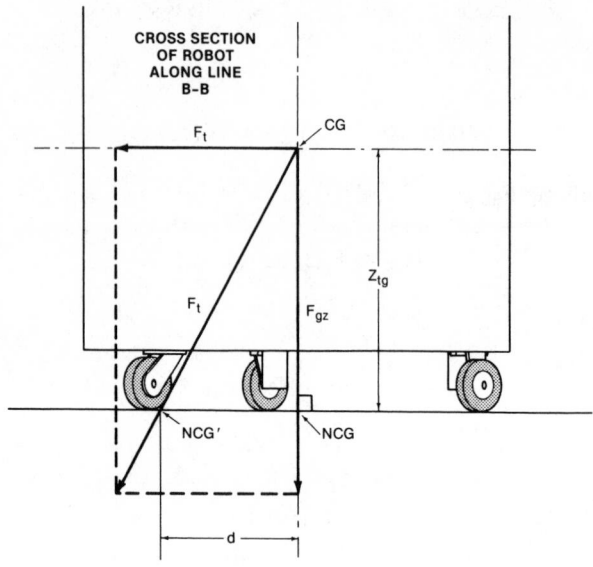

Step 2. Find the Total Lateral Force Vector.

This total lateral force is found by the vector addition of F_c, F_a, F_{gy}, & F_{gx}.

Since F_c & F_{gy} and F_a & F_{gx} act along the same lines they can be added directly.

Where,
    a is longitudinal acceleration in ft/sec^2,
    m is total mass of robot (slugs),
    r is radius of turn being executed (positive r for right-hand turn, negative r for left-hand turn),
    V is longitudinal velocity (ft/sec).

$$F_a = m \times a \text{ and } F_c = \frac{mV^2}{r}$$

Where,
    F_a is force due to acceleration,
    F_c is centrifugal force,

$$F_{tx} = F_a + F_{gx} \text{ and } F_{ty} = F_c + F_{gy}$$

246

so,

$$F_{tx} = (m \times a) + F_{gx} \qquad F_{ty} = (\frac{mV^2}{r}) + F_{gy}$$

$$F_t = \sqrt{F_{tx}^2 + F_{ty}^2} \qquad \text{(D)}$$

$$\theta_t = \arctan \left(\frac{F_{ty}}{F_{tx}} \right) \qquad \text{(E)}$$

Substituting from Equations B & C,

$$F_t = \sqrt{\left[(m \times a) + (W \times \sin \phi_g \times \cos \theta_g)\right]^2 + \left[(\frac{mV^2}{r}) + (W \times \sin \phi_g \times \sin \theta_g)\right]^2} \qquad \text{(F)}$$

The deflection distance d of NCG' is:

$$d = \frac{F_t}{F_{gz}} (Z_{cg})$$

Thus,

$$d = \frac{F_t \times Z_{cg}}{W \times \cos \phi_g} \qquad \text{(G)}$$

Equation 3-2 (H, I & J). Finding the Total Lateral Force F_t and NCG' Deflection Distance d (Step 2 - continued)

Step 2 (continued)

From equation (F) and substituting,

$$m = \frac{W}{32} \text{ (at sea level and British measure)}$$

thus,

$$F_t = W \sqrt{\left[\frac{a}{32} + (\sin \phi_g \times \cos \theta_g)\right]^2 + \left[\frac{V^2}{32r} + (\sin \phi_g \times \sin \theta_g)\right]^2} \qquad \text{(H)}$$

Substituting F_t into equation G, and canceling weight,

247

$$d = \frac{Z_{CG}}{\cos \phi_g} \sqrt{\left[\frac{a}{32} + (\sin \phi_g \times \cos \theta_g)\right]^2 + \left[\frac{V^2}{32r} + (\sin \phi_g \times \sin \theta_g)\right]^2} \quad (I)$$

and substituting for F_{ty} & F_{tx} in Equation E,

$$\theta_t = \arctan \left[\frac{\frac{V^2}{r} + (\sin \phi_g \times \sin \theta_g)}{a + (\sin \phi_g \times \sin \theta_g)}\right] \quad (J)$$

Notice that weight (and mass) is entirely absent in the final equations.

Equation 3-2 (K). Determine Stability (Step 3)

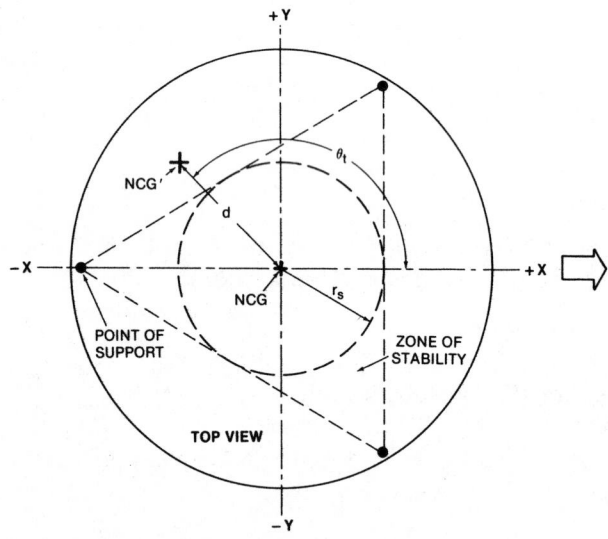

Step 3. Determine Stability.

The robot will be stable (not considering the possibility of "spinning-out") if the deflected position of NCG (NCG') lies within the zone of stability. Using θ_t, the distance from the NCG to the border of the zone can be found from a look-up table. If this distance is greater than d, the robot will be stable.

More simply, θ_t can be ignored if the worse case distance (r_s) from the NCG to the zone edge is compared to the deflection distance.

$$\text{If } d < r_s \text{ then the robot is stable} \qquad \text{(K)}$$

Note: If robot is carrying a load, NCG, r_s, H_{cg}, and m are the *effective* values of the combination of robot and load.

Equation 3-3 (A & B). Determining Pitch and Roll from a Simple Pendulum, Acceleration, & Velocity

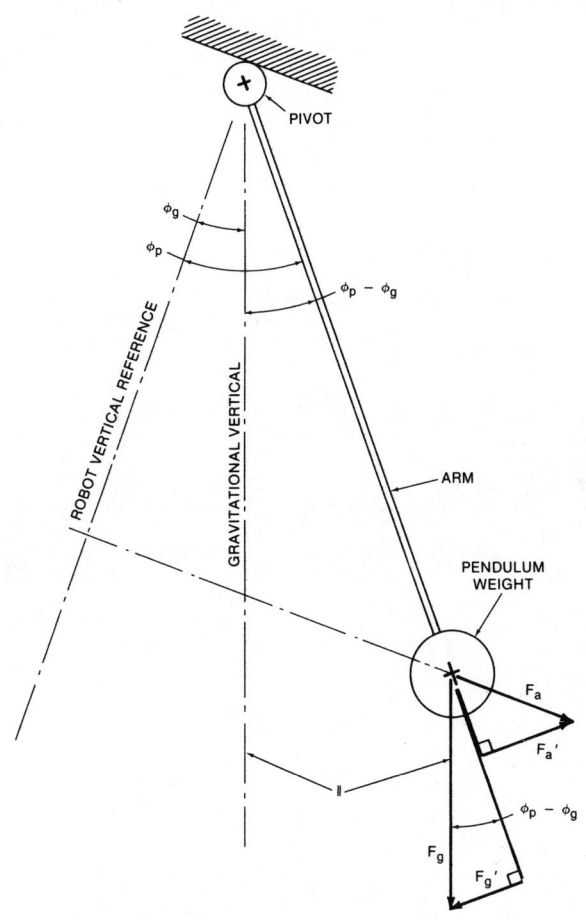

Where a is acceleration (ft/sec²):
For longitudinal (forward) direction,

$$a = \text{longitudinal acceleration}$$

For transverse direction,

$$a = \frac{V^2}{r}$$

where,
   V is velocity (ft/sec),
   r is radius of turn (ft).

F_a is acceleration force and F_g is gravitational force. The pendulum will come to rest where,

$$F_g' = F_a'$$

and,

$$F_g' = F_g \times \sin(\phi_p - \phi_g)$$

and,

$$F_a' = F_a \times \cos\phi_p$$

thus,

$$\sin(\phi_p - \phi_g) = \frac{F_a}{F_g} \times \cos\phi_p$$

or,

$$\phi_p - \phi_g = \arcsin\left(\frac{F_a}{F_g} \times \cos\phi_p\right)$$

At sea level,

$$\frac{F_a}{F_g} = \frac{m \times a}{m \times a_g} = \frac{a}{32}$$

where
   a_g is acceleration of 1G = 32 ft/sec²,
   m is mass in slugs.

Thus, for pitch

$$\phi_g = \phi_p - \arcsin\left(\frac{a}{32} \times \cos\phi_p\right) \tag{A}$$

and for roll,

$$\phi_g = \phi_p - \arcsin\left(\frac{V^2 \times \cos\phi_p}{32r}\right) \tag{B}$$

250

EQUATIONS FOR CHAPTER 4

Equation 4-1. Finding the Area

$$\text{area} = m_{(0,0)} = \frac{1}{2} \sum_{L=1}^{n} (X_L \Delta Y_L - Y_L \Delta X_L)$$

where L is a link of an n length chain.

Equations 4-2. Calculating the Moments About the System Axis

A) $A_L = (X_L \Delta Y_L - Y_L \Delta X_L)$

B) $m_{(0,0)} = \dfrac{1}{2} \sum_{L=1}^{n} A_L$

C) $m_{(1,0)} = \dfrac{1}{3} \sum_{L=1}^{n} A_L(Y_L - \dfrac{1}{2} \Delta Y_L)$

D) $m_{(0,1)} = \dfrac{1}{3} \sum_{L=1}^{n} A_L(X_L - \dfrac{1}{2} \Delta X_L)$

E) $m_{(2,0)} = \dfrac{1}{4} \sum_{L=1}^{n} A_L(X_L^2 - X_L \Delta X_L + \dfrac{1}{3} \Delta X_L^2)$

F) $m_{(0,2)} = \dfrac{1}{4} \sum_{L=1}^{n} A_L(Y_L^2 - Y_L \Delta Y_L + \dfrac{1}{3} \Delta Y_L^2)$

G) $m_{(1,1)} = \dfrac{1}{4} \sum_{L=1}^{n} A_L \left[X_L Y_L - \dfrac{1}{2} (X_L \Delta Y_L + Y_L \Delta X_L) + \dfrac{1}{3} \Delta X_L \Delta Y_L \right]$

Equation 4-3. Finding the Centers of Mass

A) $\bar{X} = \dfrac{m_{(1,0)}}{m_{(0,0)}} = \dfrac{m_{(1,0)}}{\text{area}}$

B) $\bar{Y} = \dfrac{m_{(0,1)}}{m_{(0,0)}} = \dfrac{m_{(0,1)}}{\text{area}}$

Equation 4-4. Finding the Orientation

$$\theta = \frac{1}{2} \tan^{-1} \left[\frac{2 (m_{(0,0)} m_{(1,1)} - m_{(1,0)} m_{(0,1)})}{(m_{(0,0)} m_{(2,0)} - m_{(1,0)}^2) - (m_{(0,0)} m_{(0,2)} - m_{(0,1)}^2)} \right]$$

where,

θ is angle of axis of least inertia

Equations 4-5. Calculating the Central Moments

A) $\mu_{(0,0)} = m_{(0,0)}$ (area is unchanged)
B) $\mu_{(1,0)} = 0$
C) $\mu_{(0,1)} = 0$
D) $\mu_{(2,0)} = m_{(2,0)} - \bar{x}\, m_{(1,0)}$
E) $\mu_{(0,2)} = m_{(0,2)} - \bar{y}\, m_{(0,1)}$

where,

$\mu_{(p,q)}$ is the central pth moment about the x center of mass ($\bar{x}$) and the qth moment about the y center of mass ($\bar{y}$).

Equation 4-6. Calculating the Normalized Central Moments

$$\eta_{(p,q)} = \frac{\mu_{(p,q)}}{\text{area}\left(\dfrac{p + q}{2}\right)}$$

where,

$\eta_{(p,q)}$ is the pth normalized central moment about the x axis and the qth normalized central moment about the y axis.

Equations 4-7. The Moment Invarients

A) $\phi_1 = \eta_{(2,0)} - \eta_{(0,2)}$
B) $\phi_2 = (\phi_1)^2 + 4\eta_{(1,1)}^2$
C) $\phi_3 = (\eta_{(3,0)} - 3\eta_{(1,2)})^2 + (3\eta_{(2,1)} + \eta_{(0,3)})^2$
D) $\phi_4 = (\eta_{(3,0)} + \eta_{(1,2)})^2 + (\eta_{(2,1)} + \eta_{(0,3)})^2$

Appendix B

Sample Programs
PROGRAM CG.BAS FOR DETERMINING
THE CENTER OF GRAVITY

```
REM\
***********************************************************************************\
*                                                                            *\
*            CG.BAS    Ver.    1.0    2/20/82    John M. Holland             *\
*                                                                            *\
*     This program is used to find the center of gravity of a robot. As each part is   *\
*     entered by the user, the current CG is displayed. The component data can be      *\
*     saved under a user supplied filename of up to seven letters. The filetype is always  *\
*     .COG on these files. This program is written to run under CBASIC* or to be       *\
*     compiled by CB-80* on CP/M* systems.                                     *\
*                                                                            *\
***********************************************************************************\

REM      NOTE THAT CBASIC DOES NOT REQUIRE LINE NUMBERS ON LINES                \
         THAT ARE NOT USED FOR PROGRAM BRANCHING (IE: GOTO, GOSUB).

10       MAX%=25    REM      THIS IS THE MAXIMUM NUMBER OF COMPONENTS.          \
                             IF THIS NUMBER IS CHANGED, THE PROGRAM             \
                             WILL NOT WORK WITH FILES MADE WITH OTHER           \
                             MAX% VALUE AS ARRAY SIZES WILL BE WRONG!           \

         DIM NAME$(MAX%)     REM NAMES OF COMPONENTS
         DIM WEIGHT(MAX%)    REM WEIGHTS  OF  COMPONENTS
         DIM X(MAX%)         REM X POSITION ARRAY
         DIM Y(MAX%)         REM Y POSITION ARRAY
         DIM Z(MAX%)         REM Z POSITION ARRAY

         FOR N%=1 TO 25
            PRINT ""                   REM CLEAR  SCREEN
         NEXT  N%

         PRINT "CG VER. 1.0 BY J. M. HOLLAND":PRINT

         INPUT "IS THIS A NEW DESIGN ? (Y/N):";ANSWER$
         IF UCASE$(ANSWER$)="N" THEN 20

                 INPUT "UNITS OF LENGTH ? (EG: CM, INCHES, ETC.); ";LENUNITS$
                 LENUNITS$=UCASE$(LENUNITS$)

                 INPUT "UNITS OF WEIGHT ? (EG: KG, LBS, ETC.): ";WTUNITS$
                 WTUNITS$=UCASE$(WTUNITS$)

                 GOTO40                 REM ALL DISTANCES AND WEIGHTS ARE ZERO

20       INPUT "FILE NAME ?";FILENAME$
         FILENAME$=UCASE$(FILENAME$)
         IF MATCH(".COG",FILENAME$,1)=0 THEN\
                 FILENAME$=FILENAME$+".COG"    REM ADD FILE TYPE
```

CP/M, C BASIC, and CB-80 are registered trademarks of Digital Research Corp.

```
        IF END # 1 THEN 20 REM IN CASE FILE ISNT FOUND
        OPEN FILENAME$ AS 1
        IF END # 1 THEN 30

        READ # 1;LENUNITS$ REM READ IN UNITS OF LENGTH
        READ # 1;WTUNITS$ REM READ IN UNITS OF WEIGHT
        FOR N%=1 TO MAX%
            READ # 1;NAME$(N%),WEIGHT(N%),X(N%),Y(N%),Z(N%)
        NEXT N%

30      CLOSE 1

40      REM        <<< CALCULATE CG AT THIS POINT >>>

        TOTALWT=0: XMOMENT=0: YMOMENT=0
        XMOMENT=0: ZMOMENT=0: COUNT%=0

        FOR N%=1 TO MAX%
            IF WEIGHT(N%)>0 THEN COUNT%=COUNT%+1
            TOTALWT=TOTALWT+WEIGHT(N%)
            XMOMENT=XMOMENT+X(N%)*WEIGHT(N%)
            YMOMENT=YMOMENT+Y(N%)*WEIGHT(N%)
            ZMOMENT=ZMOMENT+Z(N%)*WEIGHT(N%)
        NEXT N%

        IF TOTALWT=0 THEN 50 REM NOTHING TO CALCULATE

        PRINT: PRINT FILENAME$
        PRINT "CENTER OF GRAVITY IN ";LENUNITS$;" (X, Y, Z):"
        PRINT USING "#####.##"; \
            (XMOMENT/TOTALWT) ,(YMOMENT/TOTALWT) ,(ZMOMENT/TOTALWT)
            PRINT "TOTAL NUMBER OF COMPONENTS :";COUNT%
            PRINT "TOTAL WEIGHT OF SYSTEM        :";TOTALWT;" ";WTUNITS$

50      PRINT: INPUT "FUNCTION (? FOR INFO):";ANSWER$
        ANSWER$=UCASE$(ANSWER$)
        IF ANSWER$="?" THEN \
            PRINT "ENTER COMPONENT NUMBER (BETWEEN 1 AND ";MAX%;") OR:": \
            PRINT "CONTROL—C TO EXIT PROGRAM": \
            PRINT "L  TO LIST ALL COMPONENTS": \
            PRINT "S  TO SAVE CONFIGURATION": \
            GOTO 40

        IF ANSWER$="L"THEN 110

        IF ANSWER$="S" THEN 100

        COMP%=VAL(ANSWER$)      REM CHANGE STRING TO INTEGER COMPONENT NO.
        IF COMP%<1  THEN 40  REM INVALID COMPONENT NO.
        IF COMP%>MAX% THEN 40  REM TOO LARGE

        REM <<< PRINT COMPONENT DATA >>>

60      OLDWT=WEIGHT(COMP%)
        IF LEN(NAME$(COMP%) )=0 THEN 70
        PRINT "NAME    : ";NAME$(COMP%)

        PRINT "WEIGHT  : ";WEIGHT(COMP%);" ";WTUNITS$
        PRINT "POSITION OF COMPONENT CG IN ";LENUNITS$;" :"
        PRINT "X=";X(COMP%);"Y=";Y(COMP%);"Z=";Z(COMP%)
        PRINT
```

```
          INPUT "CHANGE ? (N=NAME, W=WEIGHT, P=POSITION, SPACE): ";ANSWER$
          ANSWER$=UCASE$(ANSWER$)
          IF ANSWER$="P" THEN 90
          IF ANSWER$="W" THEN 80
          IF ANSWER $<>"N" THEN 40

70        INPUT "NAME OF COMPONENT ? (Z TO REMOVE): ";NAME$(COMP%)
          NAME$(COMP%)=UCASE$(NAME$(COMP%) )
          IF NAME$(COMP%)="Z" THEN \ REM REMOVE ENTRY
              NAME$(COMP%)=" ": WEIGHT(COMP%)=0: \
              X(COMP%)=0: Y(COMP%)=0: Z(COMP%)=0: \
              GOTO 40

80        PRINT "WEIGHT IN ";WTUNITS$;"? :";
          INPUT " "; WEIGHT(COMP%)

90        PRINT "ALL DISTANCES IN ";LENUNITS$;"."
          PRINT "TO ZERO CG → ";
          PRINT USING "X=#####.## "; —(XMOMENT—X(COMP%)*OLDWT)/WEIGHT(COMP%)
          PRINT USING "Y=#####.## "; —(YMOMENT—Y(COMP%)*OLDWT)/WEIGHT(COMP%)
          INPUT "X POSITION ? : ";X(COMP%)
          INPUT "Y POSITION ? : ";Y(COMP%)
          INPUT "Z POSITION ? : ";Z(COMP%)
          GOTO 40

100       REM <<< SAVE DATA IN .COG FILE >>>

          INPUT "FILENAME FOR SAVING THIS DATA ? : ";FILENAME$
          FILENAME$=UCASE$(FILENAME$)
              REM STRIP ANY FILE TYPE ENTRY
          IF MATCH(".",FILENAME$,1)>0 THEN \
              FILENAME$=LEFT$(FILENAME$,MATCH(".",FILENAME$,1)—1)
          FILENAME$=FILENAME$+".COG"       REM ADD FILE TYPE

          CREATE FILENAME$ AS 1

          PRINT # 1; LENUNITS$
          PRINT # 1; WTUNITS$

          FOR N%=1 TO MAX%
              PRINT # 1;NAME$(N%),WEIGHT(N%),X(N%),Y(N%),Z(N%)
          NEXT N%

          CLOSE 1

          GOTO 40

110       REM <<<LIST ALL COMPONENTS >>>

          PRINT

          PRINT "COMP.    WEIGHT         X         Y         Z         NAME"
          FORMAT$="##     #####.##     #####.##  #####.##  #####.##         "

          FOR N%=1 TO MAX%
              IF LEN(NAME$(N%) )=0 THEN 120
              PRINT USING FORMAT$; N%,WEIGHT(N%),X(N%),Y(N%),Z(N%);
              PRINT NAME$(N%)
120       NEXT N%
          GOTO 40

          END
```

B>crun2 cg

CRUN VER 2.07P

SAMPLE RUN OF PROGRAM CG.BAS
(OPERATOR ENTRYS ARE UNDERLINED)

CG VER. 1.0 BY J. M. HOLLAND

IS THIS A NEW DESIGN ? (Y/N): Y
UNITS OF LENGTH ? (EG: CM, INCHES, ETC.): INCHES
UNITS OF WEIGHT ? (EG: KG, LBS, ETC.) : LBS

FUNCTION (? FOR INFO): ?
ENTER COMPONENT NUMBER (BETWEEN 1 AND 25) OR: } GET
CONTROL-C TO EXIT PROGRAM   } INSTRUCTIONS
L TO LIST ALL COMPONENTS
S TO SAVE CONFIGURATION

FUNCTION (? FOR INFO): 1
NAME OF COMPONENT ? (Z TO REMOVE): BASE PLATE
WEIGHT IN LBS? : 12.5
ALL DISTANCES IN INCHES.   } ENTER 1ST
TO ZERO CG → X=   0.00   Y=   0.00   } COMPONENT
X POSITION ? : 0
Y POSITION ? : 0
Z POSITION ? : 5

CENTER OF GRAVITY IN INCHES (X, Y, Z):
    0.00   0.00   5.00
TOTAL NUMBER OF COMPONENTS : 1
TOTAL WEIGHT OF SYSTEM   : 12.5 LBS

FUNCTION (? FOR INFO): 2
NAME OF COMPONENT ? (Z TO REMOVE): DRIVE MOTOR
WEIGHT IN LBS? : 12.3   } ENTER 2ND
ALL DISTANCES IN INCHES.   } COMPONENT
TO ZERO CG → X=   0.00   Y=   0.00
X POSITION ? : 4.5
Y POSITION ? : 5.7
Z POSITION ? : 8.25

CENTER OF GRAVITY IN INCHES (X, Y, Z):
    2.23   2.83   6.61
TOTAL NUMBER OF COMPONENTS : 2
TOTAL WEIGHT OF SYSTEM   : 24.8 LBS

FUNCTION (? FOR INFO): 3
NAME OF COMPONENT ? (Z TO REMOVE): STEERING MOTOR
WEIGHT IN LBS? : 3.3   } ENTER 3RD
ALL DISTANCES IN INCHES.   } COMPONENT
TO ZERO CG → X=   -16.77 Y=   -21.25
X POSITION ? : -8.4
Y POSITION ? : -12.5
Z POSITION ? : 7.7

CENTER OF GRAVITY IN INCHES (X, Y, Z):
    0.98   1.03   6.74
TOTAL NUMBER OF COMPONENTS : 3
TOTAL WEIGHT OF SYSTEM   : 28.1 LBS

```
FUNCTION (? FOR INFO): L

COMP.    WEIGHT    X       Y        Z       NAME
  1       12.50    0.00    0.00     5.00    BASE PLATE              ⎫   LIST ALL
  2       12.30    4.50    5.70     8.25    DRIVE MOTOR             ⎬   DATA
  3        3.30   -8.40  -12.50     7.70    STEERING MOTOR          ⎭   ENTERED

CENTER OF GRAVITY IN INCHES (X, Y, Z):
    0.98   1.03   6.74
TOTAL NUMBER OF COMPONENTS : 3
TOTAL WEIGHT OF SYSTEM    : 28.1 LBS

FUNCTION (? FOR INFO): 2
NAME   : DRIVE MOTOR
WEIGHT : 12.3 LBS
POSITION OF COMPONENT CG IN INCHES :
X= 4.5   Y= 5.7   Z= 8.25                                          ⎫
                                                                  ⎪   REPOSITION
CHANGE ? (N=NAME, W=WEIGHT, P=POSITION, SPACE): P                  ⎬   2ND
ALL DISTANCES IN INCHES.                                          ⎪   COMPONENT
TO ZERO CG → X=    2.25   Y=   3.35                               ⎪   TO CORRECT
X POSITION ? : 2.1                                                ⎪   CG
Y POSITION ? : 3.05                                               ⎭
Z POSITION ? : 8.25

CENTER OF GRAVITY IN INCHES (X, Y, Z):
   -0.07  -0.13   6.74
TOTAL NUMBER OF COMPONENTS : 3
TOTAL WEIGHT OF SYSTEM    : 28.1  LBS
                                                                      SAVE DATA
FUNCTION (? FOR INFO): S                                          ⎫   UNDER
FILENAME FOR SAVING THIS DATA ? : SAMPLE                          ⎬   FILENAME
                                                                 ⎭   SAMPLE.COG
SAMPLE.COG
CENTER OF GRAVITY IN INCHES (X, Y, Z):
   -0.07  -0.13   6.74
TOTAL NUMBER OF COMPONENTS : 3
TOTAL WEIGHT OF SYSTEM    : 28.1  LBS

FUNCTION (? FOR INFO): ^C                                         ⎫   EXIT
A>                                                               ⎭   PROGRAM

  B>crun2 cg
```

Recall "SAMPLE" From Disk

```
CG  VER.  1.0  BY   J. M. HOLLAND

IS THIS A  NEW  DESIGN ?  (Y/N):  N
FILE  NAME ?  SAMPLE

SAMPLE.COG
CENTER OF GRAVITY  IN INCHES  (X, Y, Z):
   -0.07    -0.13     6.74
```

```
TOTAL NUMBER OF COMPONENTS : 3
TOTAL WEIGHT OF SYSTEM    : 28.1 LBS

FUNCTION (? FOR INFO): 1

COMP.   WEIGHT    X        Y       Z      NAME
  1      12.50    0.00     0.00    5.00   BASE PLATE
  2      12.30    2.10     3.05    8.25   DRIVE MOTOR
  3       3.30   -8.40   -12.50    7.70   STEERING MOTOR

SAMPLE.COG
CENTER OF GRAVITY IN INCHES (X, Y, Z):
   -0.07    -0.13    6.74
TOTAL NUMBER OF COMPONENTS : 3
TOTAL WEIGHT OF SYSTEM        : 28.1   LBS

FUNCTION (? FOR INFO):  ^C
B>crun2 cg

CRUN VER 2.07P
```

Recall of Author's Robot Lower Torso CG Data

```
CG VER. 1.0 BY J. M. HOLLAND

IS THIS A NEW DESIGN? (Y/N): N
FILE NAME ? ROBOT1

ROBOT1.COG
CENTER OF GRAVITY IN INCHES (X, Y,Z):
   -0.13    0.06    12.85
TOTAL NUMBER OF COMPONENTS : 9
TOTAL WEIGHT OF SYSTEM        : 191.45   LBS

FUNCTION (? FOR INFO): 1

COMP.   WEIGHT    X        Y       Z      NAME
  1      63.00    0.00     0.00   13.94   BATTERIES
  2      37.95    0.00     0.00    9.00   LEGS
  3      28.00    0.00     0.00   10.09   BASE PLATE
  4       5.00    0.00     0.00   11.00   CHAINS
  5       6.50   10.80     0.00   15.19   STEERING MOTOR
  6      24.50   -3.00    -4.00   16.04   DRIVE SYSTEM
  7      10.00   -5.50     9.70   14.68   MOTOR ELECTRONICS
  8       6.50    5.00     2.00   13.68   COLLAPSE MOTOR
  9      10.00    0.00     0.00   17.50   LOWER HOUSING

ROBOT1.COG
CENTER OF GRAVITY IN INCHES (X, Y, Z):
   -0.13    0.06    12.85
TOTAL NUMBER OF COMPONENTS : 9
TOTAL WEIGHT OF SYSTEM        : 191.45   LBS

FUNCTION (? FOR INFO):
```

NOTE THAT THIS ENTIRE LIST CAN BE ENTERED AS A SINGLE COMPONENT IN ANOTHER LIST.

PROGRAM STABLE.BAS FOR DETERMINING THE LIMITS OF STABILITY

```
REM\
**************************************************************************\
*                                                                      *\
*                 STABLE.BAS    J.M. Holland    11/20/82               *\
*                               Ver.1.0                               *\
*                                                                      *\
*     This program is written to run under C BASIC or CB-80, and is used to determine  *\
*     the limits of stability for a mobile robot.  Velocity is always assumed to be    *\
*     positive, but acceleration can be either positive or negative. Refer to Equation *\
*     3.2 for variable definitions.                                   *\
*                                                                      *\
**************************************************************************\


REM      NOTE THAT C-BASIC DOES NOT REQUIRE LINE NUMBERS ON LINES \
         THAT ARE NOT USED FOR PROGRAM BRANCHING (IE: GOTO, GOSUB)

         DIM V(7)                      REM VARIABLE ARRAY
         DIM V$(7)                     REM INPUT STRING ARRAY
         DIM A$(7)
         DIM MIN(7)
         DIM MAX(7)

         DATA -60,-180,1,1,0,-32000,1
         FOR N%= 1 TO 7
             READ MIN(N%)             REM LOAD MINIMUM ARRAY
         NEXT N%

         DATA 60,180,32000,32000,32000,32000,32000
         FOR N%= 1 TO 7
         READ MAX(N%)                 REM LOAD MAXIMUM ARRAY
         NEXT N%

         FOR N%=1 TO 25
             PRINT                    REM CLEAR SCREEN
         NEXT N%

         PRINT"        STABLE VER. 1.0 J. M. HOLLAND": PRINT
         PRINT"    THIS PROGRAM DETERMINES THE WORSE CASE LIMIT"
         PRINT"    OF STABILITY FOR A ROBOT ACCELERATING AND"
         PRINT"    TURNING ON A SLOPE.  IF THE ROBOT IS RUNNING"
         PRINT"    STRAIGHT, ENTER- TURN RADIUS=1, VELOCITY=0"
         PRINT

         PRINT "PRESS [E] FOR ENGLISH UNITS."
         INPUT "PRESS [M] FOR METRIC UNITS.";ANSWER$
         ANSWER$=UCASE$(ANSWER$)
                     REM SET K TO ACCELERATION OF GRAVITY IN UNITS
         IF ANSWER$="E" THEN K=385: UNIT$="INCHES" \
         ELSE K=980: UNIT$="CM"

         V$(1)= "SLOPE STEEPNESS (-60 TO  +60 DEGREES):"
         V$(2)= "DOWN SLOPE AZIMUTH (-180 TO +180 DEGREES):"
         V$(3)= "HEIGHT OF C.G. (IN "+UNIT$+"):"
         V$(4)= "RADIUS OF TURN (IN "+UNIT$+"):"
         V$(5)= "VELOCITY (IN "+UNIT$+"/SECOND):"

         V$(6)= "ACCELERATION (IN "+UNIT$+"/SEC/SEC):"
         V$(7)= "RADIUS OF STABILITY (IN "+UNIT$+"):"
```

259

```
          A$(1)= "SLOPE (DEGREES) "
          A$(2)= "SLOPE AZIMUTH (DEGREES) "
          A$(3)= "HEIGHT OF C.G. (IN "+UNIT$+") "
          A$(4)= "RADIUS OF TURN (IN "+UNIT$+") "
          A$(5)= "VELOCITY (IN "+UNIT$+"/SECOND) "
          A$(6)= "ACCELERATION (IN "+UNIT$+"/SEC/SEC) "
          A$(7)= "RADIUS OF STABILITY (IN "+UNIT$+") "

10        PRINT " "
          PRINT "ENTER DECIMAL VALUE OR —"
          PRINT " [U] = UNKNOWN TO BE FOUND"
          PRINT " [V] = VARIABLE FOR TABLE GENERATION"
          PRINT " [B] = BACK UP ONE ENTRY"
          PRINT " [X] = ABORT ENTRY PROCESS"
          PRINT " [S] = LEAVE UNCHANGED"
          PRINT " [D] = DIAGNOSTICS"
          PRINT " "

          REM _____ INPUT  LOOP _____
          N%= 0 : VAR%= 0 : UNK%= 0 : STEPCNT%= 0
          DIAG%=0
20        N%= N%+1                        REM ENTRY LOOP
30         IF N%<1 THEN N%=1              REM LIMIT FOR B COMMAND
           IF N%>7 THEN 50
          PRINT V$(N%);V(N%);":"; REM PRINT PROMPT & PRESENT VALUE
          INPUT ANSWER$              REM INPUT VARIABLES
          ANSWER$=UCASE$(ANSWER$)
           IF ANSWER$= "D" THEN DIAG%=1: GOTO 30 REM TURN ON DIAG.
           IF  ANSWER$= "S" THEN 20      REM LEAVE UNCHANGED
           IF ANSWER$= "X" THEN 10       REM ABORT
           IF ANSWER$= "B" THEN N%=N%-1: GOTO 30
           IF ANSWER$< >"U" THEN 35
            IF UNK%< >0THEN PRINT "ONLY 1 UNKNOWN ALLOWED":\
          GOTO 30                         REM REENTER
            V(N%)= MIN(N%)
            UNK%=N%: GOTO 20              REM SET UNKNOWN REF.PNTR.
35         IF ANSWER$< >"V" THEN 40
            IF VAR%< >0 THEN PRINT "ONLY 1 VARIABLE ALLOWED":\
          GOTO 30                         REM  REENTER
            VAR%=N%                       REM SET VARIABLE PNTR.
            INPUT "STARTING VALUE:"; V(N%)
            INPUT "   STEP   SIZE:"; STEPSIZ
            INPUT "NUMBER OF STEPS:"; STEPCNT%
            GOTO 20
40         IF VAL(ANSWER$)<MIN(N%) THEN \    REM TOO SMALL
           PRINT "VALUE TOO LOW": GOTO 30
           IF VAL(ANSWER$)>MAX(N%) THEN\    REM TOO LARGE
           PRINT "VALUE TOO HIGH": GOTO 30
           V(N%)=VAL(ANSWER$)
           GOTO 20
50         IF UNK%=0 THEN \
           PRINT "MUST HAVE 1 UNKNOWN — REENTER":\

            GOTO 10
           PRINT
           REM _____ END OF INPUT _____
           REM _____ VARIABLE STEP LOOP _____
60         IF VAR%>0 THEN PRINT "FOR ";A$(VAR%);"= ";V(VAR%)

           GOSUB 100              REM FIND UNKNOWN
           IF STEPCNT%<=1 THEN 70
```

260

```
        V(VAR%)=V(VAR%)+STEPSIZ
        STEPCNT%=STEPCNT%-1 : GOTO 60
70      INPUT "ENTER CONTROL-C TO EXIT, SPACE TO CONTINUE:";\
                    ANSWER$
        GOTO 10


100     REM _____ SOLVE FOR UNKNOWN _____
        REM ____ BY USE OF SUCCESSIVE APPROXIMATIONS _____

        V(UNK%)=MIN(UNK%) REM INITIALIZE UNKNOWN TO MIN. LIMIT
        APPROX= (MAX(UNK%) -MIN(UNK%) )/4  REM SET APPROX STEP
        GOSUB 200
110     LAST%=STABLE%                       REM SAVE LAST STABILITY FLAG
        V(UNK%)=V(UNK%)+APPROX
        IF V(UNK%)<=MAX(UNK%) THEN 120
            PRINT "ROBOT IS ";
            IF STABLE%=0 THEN PRINT "UN";
            PRINT "STABLE THRU ";A$(UNK%);"RANGE (";\
                    MIN(UNK%);" TO ";MAX(UNK%);")."
        GOTO 195  REM DONT LOOK FOR SECOND SOLUTION
120     IF V(UNK%)<MIN(UNK%) THEN V(UNK%)=MIN(UNK%)
        GOSUB 200                           REM DID STABILITY CHANGE?
        IF LAST% < > STABLE% THEN 150  REM CHANGED
        GOTO 110                            REM DIDNT CHANGE SO ADD AGAIN

150     APPROX=-APPROX/2
        IF ABS(APPROX)>=1 THEN 110 REM CONTINUE APPROXIMATIONS
        PRINT "STABILITY LIMIT IS AT ";A$(UNK%);"= ";V(UNK%);
        IF DIAG%=1 THEN PRINT REM CARRIAGE RETURN IN DIAGNOSTICS

170     FIRST=V(UNK%)

        REM NOW LOOK FOR A SECOND SOLUTION
        V(UNK%)=MAX(UNK%) REM INITIALIZE UNKNOWN TO MAX LIMIT
        APPROX= -(MAX(UNK%)-MIN(UNK%) )/4 REM SET APPROX STEP
        GOSUB 200
180     LAST%=STABLE%                       REM SAVE LAST STABILITY FLAG
        V(UNK%)=V(UNK%)+APPROX
        IF V(UNK%)<MIN(UNK%) THEN PRINT : GOTO 195
185     IF V(UNK%)>MAX(UNK%) THEN V(UNK%)=MAX(UNK%)
        GOSUB 200                           REM DID STABILITY CHANGE?
        IF LAST% < > STABLE% THEN 190  REM CHANGED
        GOTO 180                            REM DIDNT CHANGE SO ADD AGAIN

190     APPROX=-APPROX/2
        IF ABS(APPROX)>=1 THEN 180 REM CONTINUE APPROXIMATIONS
        IF ABS(FIRST-V(UNK%))<2 THEN PRINT : GOTO 195 REM SAME
        PRINT " AND ";V(UNK%)
195     RETURN

200     REM _____ MAIN EQUATION _____
        SLOPE=(V(1)*3.1416)/180 REM CONVERT TO RADIANS
        AZIMUTH=(V(2)*3.1416)/180

        DX = (V(6)/K)+(SIN(SLOPE)*COS(AZIMUTH))
        DY = (V(5)^2/(K*V(4)))+(SIN(SLOPE)*SIN(AZIMUTH))
        MAG =(V(3)/COS(SLOPE))

        D = MAG*SQR(((ABS(DX))^2 + ((ABS(DY))^2))

        IF ABS(D) < V(7) THEN STABLE%=1 ELSE STABLE%=0
```

```
IF DIAG%=1 THEN PRINT \
"UNK=";V(UNK%);" DX=";DX;" DY=";DY;" D=";D;" S=";STABLE%
RETURN

END
```

SAMPLE RUN OF STABLE.BAS

```
            STABLE   VER. 1.0   J.M. HOLLAND

    THIS PROGRAM DETERMINES THE WORSE CASE LIMIT
    OF STABILITY FOR A ROBOT ACCELERATING AND
    TURNING ON A SLOPE. IF THE ROBOT IS RUNNING
    STRAIGHT, ENTER— TURN RADIUS=1. VELOCITY=0*

    PRESS [E] FOR ENGLISH UNITS.
    PRESS [M] FOR METRIC UNITS. M

    ENTER DECIMAL VALUE OR —
        [U] = UNKNOWN TO BE FOUND
        [V] = VARIABLE FOR TABLE GENERATION
        [B] = BACK UP ONE ENTRY
        [X] = ABORT ENTRY PROCESS
        [S] = LEAVE UNCHANGED
        [D] = DIAGNOSTICS                                    CURRENT VALUE

    SLOPE STEEPNESS ( -60 TO  +60 DEGREES): 0 :? U          UNKNOWN
    DOWN SLOPE AZIMUTH (-180 TO +180 DEGREES): 0 :? 0
    HEIGHT OF C.G. (IN CM): 0 :? 40
    RADIUS OF TURN (IN CM): 0 :? 1    } NO TURN*
    VELOCITY (IN CM/SECOND): 0 :? 0
    ACCELERATION (IN CM/SEC/SEC): 0 :? V          VARIABLE
    STARTING VALUE: 0
            STEP  SIZE: 200
    NUMBER OF STEPS: 8
    RADIUS OF STABILITY (IN CM): 0 :? 40

    FOR ACCELERATION (IN CM/SEC/SEC) = 0
    STABILITY LIMIT IS AT SLOPE (DEGREES) = -44  AND  44
    FOR ACCELERATION (IN CM/SEC/SEC) = 200
    STABILITY LIMIT IS AT SLOPE (DEGREES) = -53  AND  36
    FOR ACCELERATION (IN CM/SEC/SEC) = 400
    STABILITY LIMIT IS AT SLOPE (DEGREES) = 29
    FOR ACCELERATION (IN CM/SEC/SEC) = 600
    STABILITY LIMIT IS AT SLOPE (DEGREES) = 20
    FOR ACCELERATION (IN CM/SEC/SEC) = 800
    STABILITY LIMIT IS AT SLOPE (DEGREES) = 10
    FOR ACCELERATION (IN CM/SEC/SEC) = 1000
    STABILITY LIMIT IS AT SLOPE (DEGREES) = -1
    FOR ACCELERATION (IN CM/SEC/SEC) = 1200
    STABILITY LIMIT IS AT SLOPE (DEGREES) = -14
    FOR ACCELERATION (IN CM/SEC/SEC) = 1400
    ROBOT IS UNSTABLE THRU SLOPE (DEGREES) RANGE (-60  TO  60 ).
    ENTER CONTROL-C TO EXIT, SPACE TO CONTINUE:
```
                                                            RESULTS

Index

W

X

READER SERVICE CARD

   To better serve you, the reader, please take a moment to fill out this card, or a copy of it, for us. Not only will you be kept up to date on the Blacksburg Series books, but as an extra bonus, **we will randomly select five cards every month, from all of the cards sent to us during the previous month. The names that are drawn will win, absolutely free, a book from the Blacksburg Continuing Education Series.** Therefore, make sure to indicate your choice in the space provided below. For a complete listing of all the books to choose from, refer to the inside front cover of this book. Please, one card per person. Give everyone a chance.

   In order to find out who has won a book in your area, call (703) 953-1861 anytime during the night or weekend. When you do call, an answering machine will let you know the monthly winners. Too good to be true? Just give us a call. Good luck.

If I win, please send me a copy of:

_____

I understand that this book will be sent to me absolutely free, if my card is selected.

   For our information, how about telling us a little about yourself. We are interested in your occupation, how and where you normally purchase books and the books that you would like to see in the Blacksburg Series. We are also interested in finding authors for the series, so if you have a book idea, write to The Blacksburg Group, Inc., P.O. Box 242, Blacksburg, VA 24060 and ask for an Author Packet. We are also interested in TRS-80, APPLE, OSI and PET BASIC programs.

My occupation is _____
I buy books through/from _____
Would you buy books through the mail? _____
I'd like to see a book about _____
Name _____
Address _____
City _____
State _____ Zip _____

MAIL TO: BOOKS, BOX 715, BLACKSBURG, VA 24060
!!!!!PLEASE PRINT!!!!!

21952